MEDIA AND COMMUNICATIONS - TECHNOLOGIES, POLICIES AND CHALLENGES

DIGITAL NATION: BROADBAND INTERNET ADOPTION IN THE UNITED STATES

MEDIA AND COMMUNICATIONS - TECHNOLOGIES, POLICIES AND CHALLENGES

Additional books in this series can be found on Nova's website under the Series tab.

Additional E-books in this series can be found on Nova's website under the E-books tab.

MEDIA AND COMMUNICATIONS - TECHNOLOGIES, POLICIES AND CHALLENGES

DIGITAL NATION: BROADBAND INTERNET ADOPTION IN THE UNITED STATES

THOMAS N. EATON
AND
ROBERT T. CULLEN
EDITORS

Nova Science Publishers, Inc.
New York

For permission to use material from this book please contact us:
Telephone 631-231-7269; Fax 631-231-8175
Web Site: http://www.novapublishers.com

Library of Congress Cataloging-in-Publication Data

Digital nation : broadband Internet adoption in the United States / editors, Thomas N. Eaton and Robert T. Cullen.
p. cm.
Includes bibliographical references and index.
ISBN 978-1-61324-556-9 (hardcover : alk. paper) 1. Telecommunication policy--United States. 2. Broadband communication systems--United States. 3. Broadband communication systems--Government policy--United States. 4. Internet users--United States. I. Eaton, Thomas N. II. Cullen, Robert T.
HE7781.D54 2011
384.3'30973--dc23

2011014473

Published by Nova Science Publishers, Inc. † New York

CONTENTS

PREFACE

Household use of broadband internet service has risen dramatically during the first decade of the 21st century as the internet has become integral to the lives of most Americans. Nonetheless, not everyone uses broadband internet, either by choice or because they lack access. This book explores the studies that have shown that broadband internet use varies significantly across households of different socio-economic backgrounds and in different geographic locations. While broadband internet access rose between 2007 and 2009 for most demographic groups and geographic areas, persistent differences in levels and growth rates remained.

Chapter 1- Household use of broadband Internet service has risen dramatically during the first decade of the 21st century as the Internet has become integral to the lives of most Americans. Nonetheless, not everyone uses broadband Internet — either by choice or because they lack access. This report, prepared jointly by the Commerce Department's Economics and Statistics Administration (ESA) and National Telecommunications and Information Administration (NTIA), explores differences in broadband Internet use among households.

Chapter 2- The new millennium has witnessed a rapid rise in broadband Internet connections in the United States. While the growth rate of total Internet connections at home slowed somewhat after 2001, the number of household broadband connections grew rapidly over the last decade. As of October 2010, the nationwide broadband adoption rate equaled 68.2 percent of households, up from 63.5 percent one year earlier. Overall, the growth of broadband Internet adoption in the United States stems from two sources: (1) supplanting dial-up services in many households, and (2) expanding the market for household Internet connections.

Chapter 3- The FCC conducted a survey of 5,005 Americans in October and November 2009 in an effort to understand the state of broadband adoption and use, as well as barriers facing those who do not have broadband at home. The main findings are:

- 78 percent of adults are Internet users, whether that means broadband, dial-up, access from home or access from someplace other than home.
- 74 percent of adults have access at home.
- 67 percent of U.S. households contain a broadband user who accesses the service at home.
- 65 percent of adults are broadband adopters. The discrepancy of two percentage points between household and individual home use is because some survey respondents are nonbroadband users but live with someone who, at home, is.
- 6 percent of Americans use dial-up Internet connections as their main form of home access.
- 6 percent are Internet users but do not use it from home; they access the Internet from places such as work, the library or community centers.

In: Digital Nation: Broadband Internet Adoption… ISBN: 978-1-61324-556-9
Editors: Thomas N. Eaton et al.

Chapter 1

EXPLORING THE DIGITAL NATION: HOME BROADBAND INTERNET ADOPTION IN THE UNITED STATES

National Telecommunications and Information Administration

FOREWORD

The Internet Age is here. The effective use of this technology and all that it can provide is a key to success for businesses and individuals. Knowing this, the Obama Administration seeks to ensure that all Americans have affordable access to broadband Internet services. Accomplishing that goal, however, requires a set of facts about Internet use that can underpin and guide this policy objective.

In *Exploring the Digital Nation: Home Broadband Internet Adoption in the United States,* the Commerce Department fulfills its promise to provide authoritative, nationally-comprehensive data on access to the Internet throughout the United States. This new study follows the February 2010 NTIA research preview, *Digital Nation: 21st Century America's Progress Toward Universal Broadband Internet Access.* Both studies draw on the Census Bureau's Current Population Survey Internet Use Supplement, a survey of approximately 54,000 households conducted over one week in October 2009.

The Census data show increases in adoption of broadband services at home over time for virtually all demographic groups. The data also reveal that demographic disparities among groups have tended to persist. Persons with high incomes, those who are younger, Asians and Whites, the more highly-educated, married couples, and the employed tend to have higher rates of broadband use at home. Conversely, persons with low incomes, seniors, minorities, the less-educated, non-family households, and the non- employed tend to lag behind other groups in home broadband use. The new study takes the analysis to another level.

This report presents the most accurate statistical profile of U.S. broadband Internet adoption currently available. The report features new analysis of "adoption gaps," i.e., the differences in average broadband Internet adoption at home among different groups after controlling for demographic and geographic factors. There are certain groups in the population that have lower adoption rates even after taking account of differences that typically affect broadband usage. For example, the home broadband adoption gap between the lowest-income households and higher income brackets ranges from 16 to 34 percentage points, even after controlling for differences in education, age, race, ethnicity, household size, urban-rural location, foreign-born status, disability status and state of residence. The gaps between Whites and Blacks registered at 10 percentage points and between Whites and Hispanics at 14 percentage points, even after controlling for household characteristics. A similar analysis found the urban-rural gap to be 7 percentage points. A special section presents our findings on users with disabilities, who tend to be older and part of lower-income groups.

We look forward to continuing our study of this important subject in the future. Most of all, we hope that the data we make available to the public will be of use to the research community and policy- makers around the United States in their quest to understand the nature of Internet access. Through their research and ours, we hope to learn how to continue to make the benefits of this extraordinary new platform available more widely throughout the country.

Rebecca M. Blank
Under Secretary for Economic Affairs
Economics and Statistics Administration

Lawrence E. Strickling
Assistant Secretary and Administrator
National Telecommunications and Information Administration

ACKNOWLEDGMENTS

The Project Team would like to thank Mark Doms, Jane Molloy, Jane Callen, George McKittrick, Rebecca Lehrman, and Sabrina Montes of the Economics and Statistics Administration; Anna Gomez, Tom Power, Deena Shetler, Rochelle Cohen, Larry Atlas, Milton Brown, Dennis Amari, and Bart Forbes of NTIA; David Johnson of the Census Bureau; Marc Berejka and Patricia Buckley of the Office of the Secretary; Philip Weiser from the White House National Economic Council; Andrew McLaughlin and Scott Deutchman of the White House Office of Science and Technology Policy; Peter Stenberg and Christopher Chapman of the U.S. Departments of Agriculture and Education, respectively; Ceci Rouse and Matthew Magura of the Council of Economic Advisers; and John Horrigan of the Federal Communications Commission, for their contributions to this report.

EXECUTIVE SUMMARY

Household use of broadband Internet service has risen dramatically during the first decade of the 21st century as the Internet has become integral to the lives of most Americans. Nonetheless, not everyone uses broadband Internet — either by choice or because they lack access. This report, prepared jointly by the Commerce Department's Economics and Statistics Administration (ESA) and National Telecommunications and Information Administration (NTIA), explores differences in broadband Internet use among households.

Earlier studies by NTIA and others have shown that broadband Internet use varies significantly across households of different socio-economic backgrounds and in different geographic locations. This report builds on the findings of an NTIA report published earlier in 2010 showing that while broadband Internet access rose between 2007 and 2009 for most demographic groups and geographic areas, persistent differences in levels and growth rates remained (NTIA, 2010). That report found highest rates of home broadband Internet use among Asians and Whites, married couples, younger people, urban residents, people with higher incomes, and people with more education. This report expands the analysis presented in the earlier NTIA report to examine these differences in broadband Internet use in greater detail.

This report and the earlier NTIA report used data from a special 2009 supplement to the Census Bureau's Current Population Survey (CPS), which asked questions about broadband Internet use of more than 50,000 households.

The principal findings of this report are as follows:

General Broadband Internet Access

- Seven out of ten American households used the Internet in 2009. The majority of these households used broadband services to access the Internet at home. Almost one-fourth of all households, however, did not have an Internet user. (Section 3, Figure 1)

Determinants of Household Adoption of Broadband Internet

- Income and education are strongly associated with broadband Internet use at home but are not the sole determinants. (Section 4.1, Table 1; Section 4.2, Table 6)
- Broadband Internet adoption was higher among White households than among Black and Hispanic households in 2009. Differences in socio-economic attributes do not explain the entire gap in broadband Internet adoption associated with race and ethnicity. (Section 4.2, Figure 2)
- A similar pattern holds for urban and rural locations. Urban residents were more likely than their rural counterparts to adopt broadband Internet, even after accounting for socio-economic differences. (Section 4.2, Figures 3 and 4)
- Home broadband Internet use by people with disabilities lagged adoption by those with no disability. Differences in socio-economic and geographic characteristics explain a substantial portion of the adoption gap associated with disability. (Section 4.2, Figure 5)

Main Reasons for Non-Adoption of Home Broadband Internet

- Lack of need or interest, lack of affordability, lack of an adequate computer, and lack of availability were all stated as the main reason for not having home broadband Internet access. The significance of these factors, however, varied across non-users, with affordability and demand generally dominating. (Section 5, Figure 6)

- Internet non-users reported lack of need or interest as their primary reason for not having home broadband Internet access (Section 5.1, Figure 7). This group accounted for two-thirds of nonusers of home broadband Internet.
- In contrast, households that did not use the Internet specifically at home but reported using the Internet elsewhere ranked affordability as the primary deterrent to home broadband Internet use (Section 5.2, Figure 8). This group represented almost one-fourth of non-users of broadband at home.
- Affordability was also reported as the major impediment to adopting broadband Internet services in households that used dial-up services (Section 5.3, Figure 9). This group represented about one-eighth of non-users of home broadband Internet services. Lack of broadband availability was reported to be a significant factor for rural residents (Section 5.3, Table 13).
- The use of dial-up Internet service is shrinking among households that connect to the Internet from home. Dial-up users, on average, were older, had lower levels of family income and education, and were more likely to reside in rural areas. (Section 6, Tables 16 and 17)

Long-term Trends in Broadband Internet Use

- Between 2001 and 2009, broadband Internet use among households rose sevenfold, from 9% to 64% of American households utilizing broadband Internet. (Section 8.1, Table 23)
- Some of the demographic groups that had lower-than-average adoption rates in 2001 have since exhibited impressive gains. However, sizeable adoption gaps still remain in broadband Internet access among demographic groups defined by income, education, race, and ethnicity. (Section 8.1, Table 23)
- Geographic areas such as states, as well as urban and rural locations, have experienced significant growth in home broadband Internet use between 2001 and 2009. Significant gaps in adoption still persist among the states, some regions, and between urban and rural locations. (Section 8.2, Table 24; Section 8.3, Table 25)

SECTION 1. INTRODUCTION

The Internet has revolutionized the social and economic environment in which we live by providing an alternative or supplemental channel for communication, gathering and disseminating information, entertainment, commerce, and education. Household use of high-speed, or broadband, Internet services has risen dramatically during this decade which demonstrates the key role the Internet plays in the everyday lives of many individuals. Nonetheless, not everyone either uses the Internet or has access to it. This report, prepared jointly by the Commerce Department's Economics and Statistics Administration (ESA) *and* National Telecommunications and Information Administration (NTIA), seeks to explain differences in broadband Internet use among households.

Despite dramatic growth in recent years, broadband Internet use varies significantly between households with different socio-economic, demographic, and geographic characteristics. A number of recent studies, using data from different surveys, have shown this pattern. For example, a report published by NTIA earlier in 2010 using Census data found that broadband Internet access rose between 2007 and 2009 for most demographic groups defined by income, education, age, race, employment status, household type, and gender. Despite these gains, the report found the highest rates of broadband Internet adoption in 2009 among Asians and Whites, married couples, younger people, urban residents, people with higher incomes, and people with more education. The report also found that the primary reasons given by survey respondents for not having broadband Internet at home were related to affordability, demand, and availability. The Federal Communications Commission (FCC), using data from an FCC survey conducted in 2009, found that people with more education and higher income exhibit higher rates of broadband Internet use (Horrigan, 2010). They also found that Blacks and Hispanics, as well as senior citizens, lag behind in broadband Internet adoption. The Pew Internet Project, using data from a 2009 Pew survey, found that groups with historically lower broadband Internet use—including households with incomes less than $30,000, older people (50 and above), adults with only a high school degree, and rural Americans—exhibited the greatest growth in broadband Internet adoption between 2008 and 2009 (Horrigan, 2009). The most recent 2010 data from the Pew Internet Project show that Blacks experienced impressive growth in broadband Internet adoption between 2009 and 2010, while most other demographic groups experienced either moderate or no growth (Smith, 2010).

This report builds on the findings of the NTIA report published earlier in 2010 in order to more fully explore the differences in broadband Internet use among households with different characteristics. In addition, this report analyzes the main reasons provided by households for non-adoption, characteristics associated with a lag in technology adoption, and long-term growth in home broadband Internet use across population subgroups and geographic locations. The next section describes the data and methodology employed in this study. Section 3 looks at the pattern of household Internet use in 2007 and 2009. Section 4 shows how demographic characteristics and geographic location of households are associated with home broadband Internet adoption. Section 5 analyzes the main reasons provided by households for non-adoption. Section 6 analyzes who lags in technology adoption by comparing users of dial-up Internet services with users of broadband Internet services. Section 7 looks at broadband Internet adoption by people with disabilities. Finally, Section 8 studies long-term changes in home broadband and Internet use by comparing the most recent data from 2009 with that from 2001, and Section 9 provides some concluding remarks.

SECTION 2. DATA AND METHODOLOGY

This report uses data from a special supplement to the Census Bureau's Current Population Survey (CPS). The CPS is a monthly survey of a representative sample of the U.S. population, and provides data on labor force participation, income, and demographic characteristics of households. The special supplement utilized in this report, the CPS Internet Use Supplement, periodically gathers information on household Internet use. This report analyzes data from the most recent survey conducted in October 2009, the eighth such Internet survey conducted since the early 1990s. The October 2009 CPS interviewed 54,324 households. For a more detailed description of the survey, see Section Al of the Appendix.

The October 2009 Internet Use Supplement asked each household whether someone in that household used the Internet and what kind of Internet connection technology was utilized at home (the respondent was asked to choose from three options: "dial-up" telephone service, "broadband,"[1] or "something else"). The survey also asked households in which no one used the Internet or where a "dial-up" telephone service was utilized, to state their main reason for not using broadband Internet services. Using these data, one can therefore identify households and individuals who use broadband Internet at home to connect to the Internet. This report focuses on broadband Internet use at the household level, as opposed to individual level, since the decision to adopt a particular type of Internet service technology at home likely occurs at the household level.[2] One would expect a household to evaluate the cost of the technology relative to the collective benefit of the technology for all household members. The outcome of this decision-making process, comparing the costs versus the collective benefits, is likely to vary across household types.

Our sample consists of all households where the age of the head of the household is 16 or above. There are 54,280 household records in our sample, representing 119 million American households. We analyze broadband Internet use at the household level and its association with household-level characteristics. For characteristics like education, race, ethnicity, age, disability status, and foreign- born status, we use the information for the head of household. In this report we use the words "adoption," "use," "utilization," and "connectivity" interchangeably in order to indicate that a household reported using a broadband service at home to connect to the Internet.

As mentioned in the previous section, this report builds on the findings in NTIA (2010) which shows that certain population subgroups, specifically people with higher incomes, those with more education, Asians, Whites, married couples, those who are younger, and residents of urban areas exhibit the highest rates of broadband Internet use. These findings point to several areas of inquiry, particularly whether socio-economic differences among households explain the differences in broadband Internet use. For instance, is the higher rate of broadband Internet use among urban households explained by the differences in income and education between urban and rural households? Or, stated another way, how much of an urban-rural difference in broadband Internet use remains if we compare urban and rural households with similar income, education, and other key characteristics? Similarly, how much of a difference in broadband Internet use between White and Hispanic households remains if we compare adoption between White and Hispanic households that have similar income, education, geographic location, and other observed characteristics?

A simple tabulation of the data by household characteristics does not allow the researcher to answer such questions. By utilizing a regression analysis framework, we can estimate the marginal or "isolated" association between broadband Internet use and a particular household attribute. For example, the marginal effect of income on broadband Internet use can be estimated by comparing broadband Internet use among households that have different income levels but which are otherwise similar with respect to key attributes like education, race, ethnicity, age, geographic location, and other possible determinants of broadband Internet use. The regression will tell us how much the likelihood of broadband Internet use would rise for a given increase in income, holding key demographic and geographic characteristics constant.

The next section of this report (Section 3) looks at home broadband Internet usage patterns for 2007 and 2009, and Section 4 employs a regression analysis framework to analyze how much of the observed differences in broadband Internet use across households is explained by differences in socio-economic and geographic factors.

SECTION 3. BROADBAND INTERNET USE IN 2007 AND 2009

The top half of Figure 1 shows the pattern of Internet use among American households in 2009. Figure 1 shows that a significant portion of American households (64%) connected to the Internet from home utilizing a broadband Internet service. Another 5% used dialup services at home to connect to the Internet. This means that almost seven out of ten American households (69%) connected to the Internet from home in 2009.[3] Another 8% used the Internet at a location other than home, implying that more than three-fourths of all American households (77%) had at least one person who used the Internet in 2009. This was up from 71% in 2007 (the distribution for 2007 is presented in the bottom half of Figure 1).

Broadband was by far the most frequently used technology for accessing the Internet from home. The share of households subscribing to broadband Internet services rose from 51% in 2007 to 64% in 2009, implying that home broadband Internet use rose by one-fourth during the two year period. Households with a dial-up Internet service accounted for a shrinking share of Internet users — down from 11% in 2007 to 5% in 2009. The decline in dial-up users has been outweighed by the rise in broadband users, resulting in a net increase in Internet use at home.

Despite this overall growth in Internet use, it is important to realize that a significant portion of American households (36%) did not have a broadband Internet service at home. Almost one-fourth of American households (23%) did not have any Internet user in 2009.

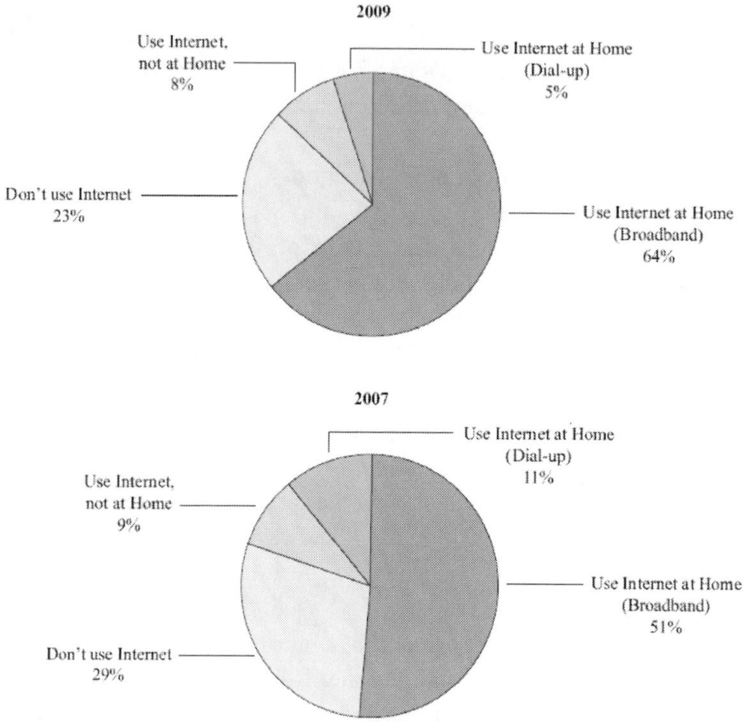

Source: U.S. Census Bureau, Current Population Survey (CPS) and CPS School Enrollment and Internet Use Supplement, October 2007 and October 2009, and ESA calculations.

Figure 1. Distribution of Internet Use by Households, 2007 and 2009.

SECTION 4. BROADBAND INTERNET USE IN 2009: DEMOGRAPHIC AND GEOGRAPHIC CHARACTERISTICS

This section will analyze how broadband Internet use varies across households of different demographic and socio-economic backgrounds and in different geographic locations. Section 4.1 looks at average broadband Internet adoption rates by household characteristics and Section 4.2 utilizes a regression analysis framework that enables us to isolate the impact of any one factor or characteristic on broadband Internet adoption. The results indicate that home broadband Internet use is more prevalent among households with higher incomes and more education as well as among Whites and urban households, and that the gaps in adoption between White and non-White households or between urban and rural households are not entirely explained by differences in socioeconomic and demographic factors.

Section 4.1. Broadband Internet Use by Household Characteristics

Table 1 shows average broadband Internet usage rates by demographic characteristics. According to Table 1, home broadband Internet use is more prevalent among households with higher incomes, more education, Asians, and Whites. For example, slightly more than a third (36%) of households with annual family incomes less than $25,000 used broadband Internet at home in 2009, compared to the majority of households with higher incomes. Slightly more than one-fourth (29%) of households headed by someone with less than a high school degree used broadband Internet at home, compared to the vast majority (85%) of their counterparts with a college degree or more. Non-Hispanic Asian households (77%) had the highest rate of broadband Internet use in 2009, followed by non-Hispanic White households (68%). Hispanic (48%) and non-Hispanic Black (49%) households lagged behind with adoption rates that were about 20 percentage points lower than their non-Hispanic White counterparts.

Table 1 also shows that broadband Internet use is strongly correlated with age, household type, and disability status. Seventy-one percent of households where the head of the household was between 16 and 44 years of age had broadband Internet at home, compared to 40% of their counterparts aged 65 years or more. The majority of married-couple families with children used broadband Internet services at home (80%), compared to about two-thirds of family households without children (68%) and half of non-family households (51%). Households headed by someone with a disability were almost half as likely as households headed by someone with no disability to have broadband Internet (38% compared to 68%).[4] Finally, foreign-born non-U.S. citizens were less likely than American citizens to utilize broadband Internet at home (51% compared to 64%).[5]

Table 2 shows average broadband Internet usage rates by geographic location of households (by urban-rural status and by size of urban area). Note that this report uses the terms "urban" and "rural" to refer to metropolitan and nonmetropolitan areas, respectively.[6] According to Table 2, broadband Internet adoption, on average, was 15 percentage points higher among urban dwellers than among their rural-area counterparts (66% versus 51%). Urban areas with populations between 2.5 million and 5 million displayed the highest broadband Internet adoption rate with 71% of households living in these areas subscribing to broadband Internet services.

Table 1. Household Broadband Internet Use by Demographic Characteristics, 2009

Household Broadband Internet Use: Percent of households connecting to the Internet at home using broadband	
All Households*	63.5
Household Income	
Less than $25,000	35.8
$25,000-$50,000	61.0
$50,000-$75,000	79.3
$75,000-$100,000	87.6
More than $100,000	94.1
Education	
Less than High School Degree	28.8
High School Degree	50.9
Some College	69.5
College Degree or more	84.5
Race and Ethnicity	
White, Non-Hispanic	68.0
Black, Non-Hispanic	49.4
Asian, Non-Hispanic	77.3
American Indian or Alaskan Native, Non-Hispanic	48.3
Hispanic	47.9
Age	
16 to 44 years	71.2
45 to 64 years	68.2
65 years and over	39.9
Gender	
Male	66.7
Female	60.2
Household Type	
Married-couple with children	79.8
Single parent (male)	60.1

Household Broadband Internet Use: Percent of households connecting to the Internet at home using broadband	
Single parent (female)	56.9
Family without children	67.7
Non-family household	50.8
Disability Status	
Has a disability	37.8
No disability	67.6
Foreign-Born Status	
Citizens (including foreign born)	64.4
Non-Citizen	51.0
Sample Size	54,280
Estimated Number of Households	119,267,400

Source: U.S. Census Bureau, Current Population Survey (CPS) and CPS School Enrollment and Internet Use Supplement, October 2009, and ESA calculations.

Note: *Sample includes all households with the head of the household at least 16 years of age. The information for the head of the household is used for education, race, ethnicity, age, gender, foreign-born status, and disability.

Table 2. Household Broadband Internet Use by Geographic Characteristics, 2009

Household Broadband Internet Use: Percent of households connecting to the Internet at home using broadband	
All Households	63.5
Urban-Rural Status	
Urban (Metropolitan)	65.9
Rural (Nonmetropolitan)	51.0
Metropolitan Area (CBSA) Size	
Under 1,000,000	63.1
1,000,000-2,499,999	66.1
2,500,000-4,999,999	70.5
5,000,000 or more	66.9
Sample Size	54,280
Estimated Number of Households	119,267,400

Source: U.S. Census Bureau, Current Population Survey (CPS) and CPS School Enrollment and Internet Use Supplement, October 2009, and ESA calculations.

The next three tables dig deeper into the data by looking at average broadband Internet usage rates by race, ethnicity, income, and education within urban and rural areas (Table 3) and then by cross tabulating broadband Internet usage data by race and income, and by race and education for urban and rural areas separately (Tables 4 and 5). Table 3 shows that rural households with lower incomes, lower levels of education, and Black rural households had particularly low broadband Internet adoption rates. Only about 28% of rural dwellers with incomes less than $25,000 had broadband Internet at home, compared to 38% of their urban counterparts and 86% of their high-income rural counterparts. A similar pattern holds for demographic groups defined by race, ethnicity, and education.

Table 3. Household Broadband Internet Use by Metropolitan Status, 'Race, arri Income, 200°

Household Broadband Internet Use: Percent of households connecting to the Internet at home using broadband		
	Urban	**Rural**
All Households	65.9	51.0
Race and Ethnicity		
Non-Hispanic White	71.2	54.2
Non-Hispanic Black	52.1	28.7
Hispanic	48.6	36.9
Household Income		
Less than $25,000	38.0	28.0
$25,000-$50,000	62.8	52.4
$50,000-$75,000	80.2	73.6
$75,000-$100,000	88.6	81.3
More than $100,000	94.8	86.1
Education		
Less than High School Degree	30.5	21.7
High School Degree	52.8	43.6
Some College	70.8	62.0
College Degree or more	85.5	74.7

Source: U.S. Census Bureau, Current Population Survey (CPS) and CPS School Enrollment and Internet Use Supplement, October 2009, and ESA calculations.

Table 4. Household Broadband Internet Use by Metropolitan Status, Race, and Income, 2009

Household Broadband Internet Use: Percent of households connecting to the Internet at home using broadband		
Race and Income	**Urban**	**Rural**
Non-Hispanic White		
Household Income: Less than $25,000	42.6	31.3
Household Income: $25,000-$50,000	66.3	53.8
Household Income: $50,000-$75,000	81.7	74.6
Household Income: More than $75,000	92.8	84.2
Non-Hispanic Black		
Household Income: Less than $25,000	33.1	16.8
Household Income: $25,000-$50,000	59.6	36.9
Household Income: $50,000-$75,000	76.4	61.5
Household Income: More than $75,000	88.2	71.9
Hispanic		
Household Income: Less than $25,000	27.4	18.5
Household Income: $25,000-$50,000	49.3	39.6
Household Income: $50,000-$75,000	71.3	70.0
Household Income: More than $75,000	88.7	73.1

Source: U.S. Census Bureau, Current Population Survey (CPS) and CPS School Enrollment and Internet Use Supplement, October 2009, and ESA calculations.

A similar pattern is displayed in Table 5, which shows the cross-tabulated data on average broadband Internet use by race and educational attainment, and by urban-rural location. Black households headed by someone with less than a high school degree and living in rural areas exhibited the lowest level of broadband Internet use (11%). They were followed by White rural households headed by someone with less than a high school degree (23%), Black rural households headed by someone with a high school degree (24%), Hispanic households headed by someone with less than a high school degree (both urban and rural) (26% and 25%, respectively), and Black urban households headed by someone with less than a high school degree (27%).

Table 5. Household Broadband Internet Use by Metropolitan Status, Race and Education, 2009

Household Broadband Internet Use: Percent of households connecting to the Internet at home using broadband		
Race and Education	**Urban**	**Rural**
Non-Hispanic White		
Less than High School Degree	34.4	23.3
High School Degree	56.6	46.4
Some College	74.1	64.3
College Degree or more	86.9	75.6
Non-Hispanic Black		
Less than High School Degree	26.9	10.7
High School Degree	42.4	24.0
Some College	57.0	43.0
College Degree or more	76.9	55.8
Hispanic		
Less than High School Degree	26.0	25.0
High School Degree	45.4	30.8
Some College	67.0	56.0
College Degree or more	77.9	68.8

Source: U.S. Census Bureau, Current Population Survey (CPS) and CPS School Enrollment and Internet Use Supplement, October 2009, and ESA calculations.

Section 4.2. Marginal Effects of Demographic and Geographic Characteristics on the Likelihood that a Household Uses Broadband Internet at Home

The finding that socio-economic characteristics, as well as race, ethnicity, and geographic location are highly correlated with adoption of home broadband Internet services has important implications. These household attributes are themselves correlated with each other. For instance, income and education are likely to be higher in urban areas if employment opportunities requiring high levels of skills and specialization are disproportionately located in urban areas. As a result, it is not clear from the tabulations we have seen so far how much of the urban-rural gap in adoption is driven by differences in income and education between urban and rural residents. The same issue applies for race and ethnicity, that is, looking at average adoption levels by race and ethnicity does not tell us how much of the adoption gap associated with race and ethnicity is explained by differences in socio-economic factors.

The rest of this section will utilize a regression analysis framework that estimates the impact of multiple factors together on the probability that a household adopts broadband Internet services at home. The results allow us to isolate or distinguish the effect of any one factor while holding all other factors constant. We refer to these results as the marginal effect of selected demographic and geographic characteristics on household broadband Internet use. The factors that we control for in this analysis include household income, education, age, race, ethnicity, foreign-born status, household size (total number of persons in household), disability status, and geographic location (urban-rural location and state). Note that the CPS data do not provide information on broadband Internet availability and price in a household's immediate location, which is why we are unable to directly control for these factors. Both price and availability are important determinants of adoption. The regression analysis, however, accounts for a household's geographic location (urban versus rural location, the size of the urban area a household lives in, and state)[7] As a result, these household geographic characteristics would capture some of the variation in broadband Internet price and availability along these geographic dimensions.

The full set of regression results from this analysis is presented in Section A3 of the Appendix. Table 6 presents the estimated marginal effects of selected demographic and geographic factors on the probability of broadband Internet adoption at home. The marginal effect of a particular household characteristic, for instance, the impact of living in an urban location, is the isolated effect of an urban location on the likelihood of broadband Internet use, after holding constant the above mentioned characteristics. In other words, the marginal effect of living in an urban location is the gap in average broadband Internet adoption between urban and rural households, after accounting for differences between urban and rural households in income, education, age, race, ethnicity, household size, foreign-born status, disability status, and state of residence.

According to the results in Table 6, the likelihood of broadband Internet adoption among households with incomes between $25,000 and $50,000 is 16 percentage points higher than that among households with incomes less than $25,000, after accounting for differences in other characteristics (education, age, race, ethnicity, household size, foreign-born status, disability status, urban-rural status, and state of residence). Note that Table 1 showed average broadband Internet usage rates of 61% and 36% for these two groups, implying an adoption gap of 25 percentage points before controlling for these other factors. This means that differences in other characteristics like education, race, ethnicity, age, geographic location, household size, foreign-born status, and disability explain some of the differences in broadband Internet adoption between these two groups. Therefore, the adoption gap declines from 25 percentage points to 16 percentage points once we account for these attributes.

Table 6. Marginal Effects of Selected Demographic and Geographic Characteristics on the Likelihood that a Household Uses Broadband Internet at Home, 2009

Adoption Gap: Difference in average broadband Internet adoption after controlling for demographic and geographic factors	
Household Characteristic	**Adoption Gap (Percentage point)**
Household Income	
Gap between households with incomes $25,000 to $50,000 and households with incomes less than $25,000	16
Gap between households with incomes $50,000 to $75,000 and households with incomes less than $25,000	27
Gap between households with incomes $75,000 to $100,000 and households with incomes less than $25,000	31
Gap between households with incomes more than $100,000 and households with incomes less than $25,000	34
Education	
Gap between those with a high school degree and those with less than high school degree	11
Gap between those with some college and those with less than high school degree	23
Gap between those with college degree or more and those with less than high school degree	29
Race and Ethnicity	
Gap between Non-Hispanic White and Non-Hispanic Black	10
Gap between Non-Hispanic White and Hispanic	14
Gap between Non-Hispanic White and Non-Hispanic Asian	0

Household Characteristic	Adoption Gap (Percentage point)
Gap between Non-Hispanic White and Other*	5
Urban-Rural	
Gap between urban and rural households	7
Foreign-born Status	
Gap between U.S. citizens and foreign-born non-citizens	6
Disability	
Gap between those with no disability and with disability	5
Sample Size	43,662
Estimated Number of Households	94,963,684

Source: U.S. Census Bureau, Current Population Survey (CPS) and CPS School Enrollment and Internet Use Supplement, October 2009, and ESA calculations.

Note: Sample includes all households with the head of the household at least 16 years of age and with non-missing data on household income. The sample size declines from 54,280 for tabulations to 43,662 for regressions because of excluding observations with missing data on household income. Controls for age, household size and state of residence are included. See Appendix Table A3 (column 1) for the full set of regression results. *This category includes Native Americans, Hawaiian/Pacific Islanders, and those who report two or more races.

Relative to households in the lowest income category (less than $25,000), the increased likelihood of adoption, or adoption gap, is 27 percentage points for households with family incomes between $50,000 and $75,000; 31 percentage points for households with family incomes between $75,000 and $100,000; and 34 percentage points for households with family incomes exceeding $100,000. Two implications are apparent. First, income is strongly associated with broadband Internet use. The positive association between income and broadband connectivity persists even after accounting for differences in a large number of key characteristics including education, age race, ethnicity, and geography. Second, the rising effect of income diminishes as income grows.

A similar phenomenon is apparent for education. According to Table 6, the likelihood of broadband Internet use, on average, is 11 percentage points higher among households that are headed by someone with a high school degree compared to households that are headed by someone with less than a high school degree, again holding all other factors equal. The adoption gap is 23 percentage points between those with some college and those without a high school diploma, and 29 percentage points between those with at least a college degree and those without a high school diploma. As with income, this suggests that education is strongly associated with broadband Internet adoption, even after accounting for differences in income, age, race, ethnicity, and a number of other key characteristics. Households with higher levels of income and education are more likely to have the necessary resources and skills to obtain and use broadband Internet services at home.

Table 1 showed average broadband Internet usage rates of 68% for non-Hispanic White households, 49% for non-Hispanic Black households, and 48% for Hispanic households, implying that the gap in average adoption was 19 percentage points between White and Black households, and 20 percentage points between White and Hispanic households. Table 6 shows that, once we have controlled for socio-economic and geographic attributes, the White-Black adoption gap declines to 10 percentage points and the White-Hispanic adoption gap declines to 14 percentage points. Figure 2 plots these White-Black and White-Hispanic adoption gaps. Figure 2 uses two bars to display the adoption gap between any two groups of people. The bottom bar of each pair (which is also the longer bar) shows the gap in average adoption from Table 1. The top bar of each pair (which is also the shorter bar) shows the remaining unexplained adoption gap after accounting for differences in household demographic, socio-economic, and geographic characteristics (from Table 6). The remaining gap suggests that the broadband Internet adoption gap associated with race and ethnicity is not entirely explained by differences in income or other non-income attributes. Income, education, age, foreign-born status, and other demographic and geographic characteristics explain about one-half of the White-Black gap and one-fourth of the White-Hispanic gap in broadband Internet usage. As a result, a sizeable gap in adoption remains after controlling for socio-economic and geographic factors.

Table 2 showed average broadband Internet usage rates of 66% for urban households and 51% for rural households, reflecting a 15 percentage point urban-rural gap in broadband Internet usage. Our analysis shows that differences in socio-economic and demographic characteristics explain about half of this urban-rural adoption gap. In other words, an adoption gap of 7 percentage points remains between urban and rural dwellers even after controlling for differences in income, education, race, ethnicity, age, household size, foreign-born status, disability status, and state of residence. Figure 3 shows the urban-rural adoption gap.

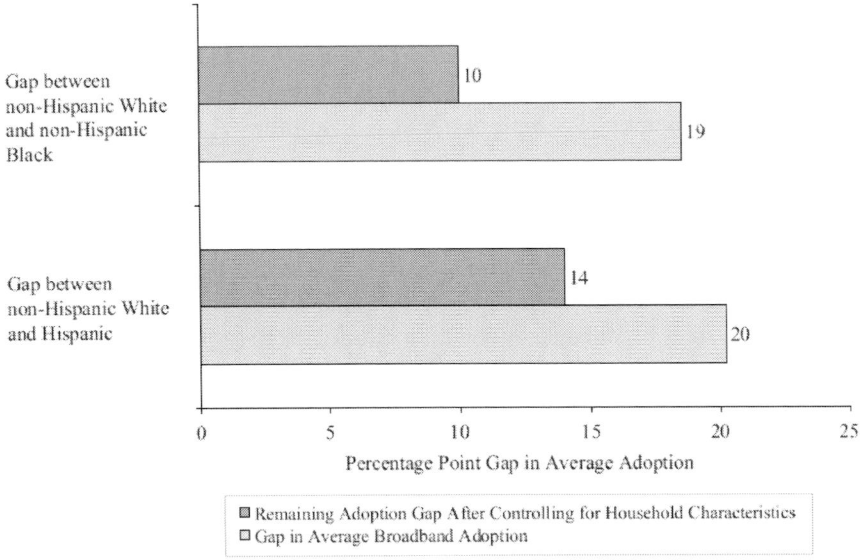

Source: U.S. Census Bureau, Current Population Survey (CPS) and CPS School Enrollment and Internet Use Supplement, October 2009, and ESA calculations.

Figure 2. Race and Ethnicity-related Gap in Broadband Internet Adoption before and after Controlling for Household Characteristics, 2009.

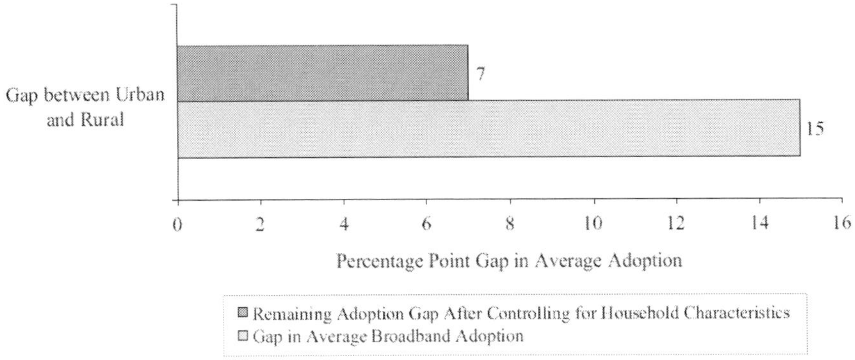

Source: U.S. Census Bureau, Current Population Survey (CPS) and CPS School Enrollment and Internet Use Supplement, October 2009, and ESA calculations.

Figure 3. Urban-Rural Gap in Broadband Internet Adoption Before and after Controlling for Household Characteristics, 2009.

Our analysis also shows that the urban-rural gap in broadband Internet adoption varies with the size of the urban area. Compared to rural residents, the likelihood of broadband Internet use is 9 percentage points higher for households residing in urban areas with populations exceeding two and half million, and 6 to 7 percentage points higher for households living in urban areas with populations less than two and half million (Appendix Table A3, Column 2). According to Table 2, the urban-rural gap in adoption, before controlling for socio-economic factors and state of residence, ranged from 12 percentage points to 20 percentage points depending on the size of the urban area. Figure 4 plots these adoption gaps by the size of the urban area. This again implies that socio-economic factors explain a substantial, but not the entire, urban-rural broadband Internet adoption gap. Broadband Internet price and availability are likely to explain some of the remaining urban-rural gap– the higher adoption rates in larger urban areas may be driven by lower prices and more availability of broadband Internet services in these areas.

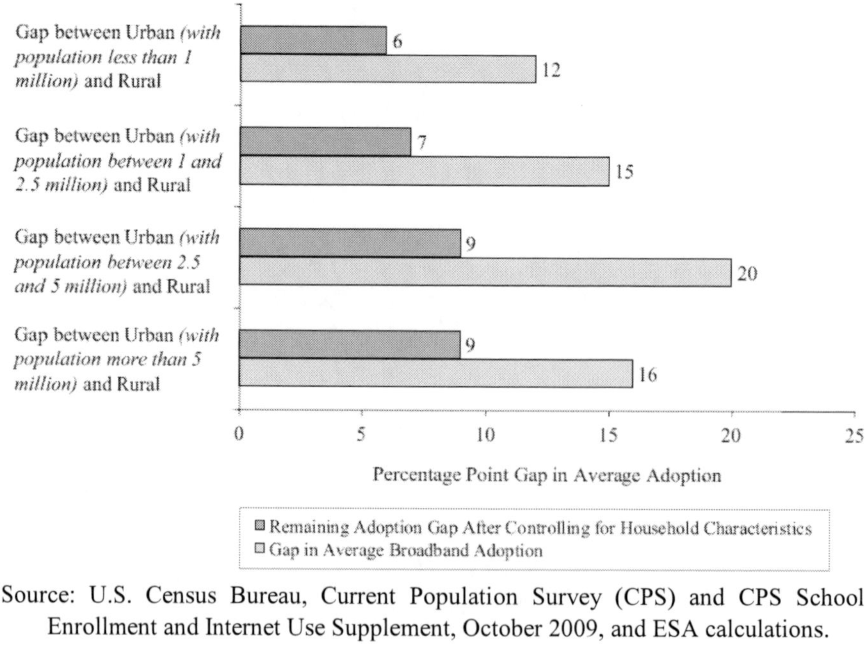

Source: U.S. Census Bureau, Current Population Survey (CPS) and CPS School Enrollment and Internet Use Supplement, October 2009, and ESA calculations.

Figure 4. Urban-Rural Gap in Broadband Internet Adoption Before and after Controlling for Household Characteristics, by Urban Area Size, 2009.

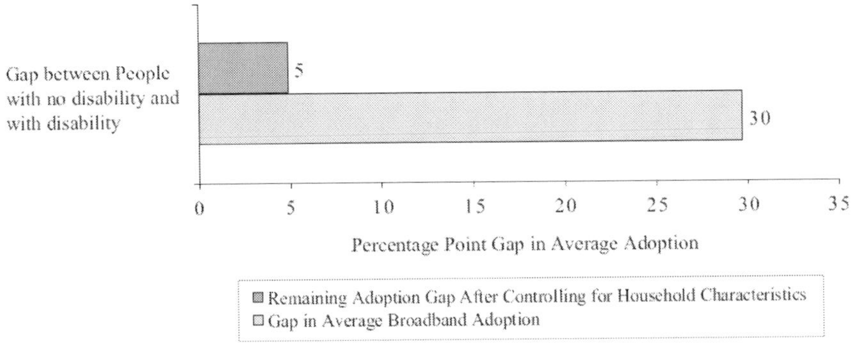

Source: U.S. Census Bureau, Current Population Survey (CPS) and Enrollment and Internet Use Supplement, October 2009, and ESA calculations.

Figure 5. Disability-related Gap in Broadband Internet Adoption Before and After Controlling for Household Characteristics, 2009.

Table 1 showed average broadband Internet adoption rates of 38% for households headed by someone with a disability and 68% for households headed by someone with no disability, implying a 30 percentage point adoption gap. The gap in adoption declines to 5 percentage points after controlling for socio-economic and geographic characteristics (Table 6), implying that the vast majority of the adoption gap associated with disability is explained by differences in these factors. This was not the case for race and ethnicity where a sizeable gap in adoption remained. Figure 5 shows the adoption gap associated with disability.

Table 6 also shows that foreign-born non-citizen households were, on average, 6 percentage points less likely than their U.S. citizen counterparts to subscribe to broadband Internet. This means that income, education, race, ethnicity and other observed characteristics explain more than half of the initial 13 percentage point gap in adoption between foreign-born non-citizens and American citizens (Table 1).

In addition, our estimates show that the likelihood of broadband Internet use increases with age up to about age 30 to 35, after which it declines as age increases (not shown here). The state indicator variable accounts for a separate state-specific effect (these are not reported) and shows that significant differences exist in broadband Internet use across states, even after accounting for differences in income, education, urban-rural status, and other household characteristics. Finally, our analysis shows that the likelihood of home broadband Internet use increases with household size.

In summary, the regression results presented in this section suggest that income and education are strongly associated with broadband Internet use at home but are not the sole determinants. Other factors, particularly race, ethnicity, and urban-rural location, are also independently associated with home broadband Internet adoption. The adoption gaps associated with race and ethnicity, or urban and rural locations, are not entirely explained by socio-economic characteristics, i.e., these gaps do not disappear after accounting for differences in income, education, and a number of other key household attributes. Socio-economic factors, however, explain a substantial portion of the adoption gap associated with disability.

As mentioned previously, the decision to adopt broadband or any other type of Internet service technology at home likely occurs at the household level after evaluating the cost of the technology relative to the collective benefit of the technology for all household members. This suggests that the decision-making process is likely to vary across household types. We looked at the marginal effects of socio-economic and geographic attributes for four different household types—married couples with children, single parents with children, family households without children, and non-family households. The association between broadband Internet use and socio-economic and geographic attributes was quite robust across different household types. A brief discussion of these results is presented in Section A4 of the Appendix.

The CPS data do not provide information on price and availability of broadband Internet in a household's immediate location, which is why we are unable to account for these factors and therefore unable to distinguish how much of the variation across socio-economic and geographic dimensions is likely driven by demand versus supply-related factors. Part of the non-adoption may result from lower demand for broadband Internet, related to affordability or cost. But some non-adoption may occur because of lack of supply or availability of broadband Internet services. For instance, are rural residents less likely to adopt because they have lower demand for broadband Internet or because broadband Internet availability is limited in their location?

The CPS Internet Use Supplement does ask households to state their main reason for not using home broadband Internet services. The next section will analyze these reasons. We will see that factors like affordability, perceived need or interest, complementary equipment, and availability all play significant roles in a household's adoption decision.

SECTION 5. MAIN REASON FOR NON-ADOPTION OF HOME BROADBAND INTERNET

This section will analyze the principal reason for not having broadband Internet access at home. The CPS asks three types of non-users to state their main reason for not using home broadband Internet services—households that do not use the Internet generally, households that do not use the Internet specifically at home but report using the Internet elsewhere, and households that use a dial-up Internet service at home. Figure 6 tabulates the responses from all households without broadband Internet access, whereas Sections 5.1-5.3 separately analyze the responses from each group in order to understand whether adoption decisions of different groups are impacted by different factors. Note that the reasons provided by households reflect their subjective opinion since a household may not have full information on pricing, availability, or the benefit of using broadband Internet. For instance, a household may believe that broadband Internet is not available in its area, but be misinformed. As a result, any comparison across households, while informative, needs to be done with caution.

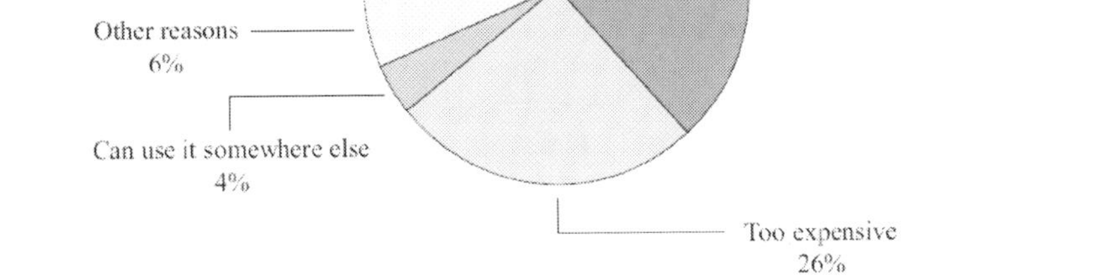

Source: U.S. Census Bureau, Current Population Survey (CPS) and CPS School Enrollment and Internet Use Supplement, October 2009, and ESA calculations.

Note: The figure does not include the categories that were reported as the main reason by less than 1% of non-users.

Figure 6. Main Reason for Non-Adoption of Home Broadband Internet, 2009.

Figure 6 shows that the most commonly cited reason for not having broadband Internet access at home was "don't need" (38%), followed by "too expensive"(26%) and "inadequate computer" (18%). The next three sections, however, will show that the relative significance of these factors varies across different types of non-users.

SECTION 5.1. AMONG INTERNET NON-USERS

Figure 1 in Section 3 showed that almost one-fourth of American households (23%) in 2009 reported that no one in those households used the Internet at any location. This section will analyze the reasons given by these households for not having broadband Internet access at home. This group accounts for 65% of all those that do not access broadband Internet at home.

Lack of need or interest was the most commonly cited reason for not using broadband Internet services at home. Figure 7 shows that 47% of households who did not use the Internet cited "don't need it—not interested" as their principal reason for not subscribing to home broadband Internet services. Another 22% cited lack of an adequate computer. Only 19% cited affordability or cost. This means that a perceived lack of value or need was a more significant factor than affordability for non-use of broadband Internet services.

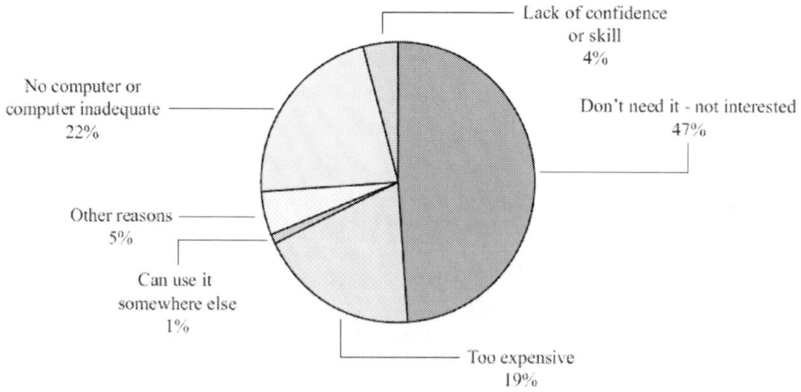

Source: U.S. Census Bureau, Current Population Survey (CPS) and CPS School Enrollment and Internet Use Supplement, October 2009, and ESA calculations.
Note: The figure does not include the categories that were reported as the main reason by less than 1% of non-users. See column 1 of Table 7 for the full set of results.

Figure 7. Main Reason Provided by Internet Non-Users, 2009.

Table 7 shows the results for the overall sample as well as for urban and rural households. The two right columns of Table 7 show that the rankings are largely unchanged for urban and rural households. A larger share of rural households than their urban counterparts, however, stated lack of need as the major reason (52% compared to 46%), while a smaller share of rural households stated affordability as the major deterrent (16% compared to 20%). Lack of availability was not a significant impediment for either group.

Table 8 shows the reasons by income categories. A lack of need or interest was the primary reason for non-use in every income group. Affordability was much more important for low-income households, however. Lack of an adequate computer played an important role for all households.

Table 9 identifies the reasons for non-use by race and ethnicity. The most important factor for all groups was need, although more White households than Black and Hispanic households gave this as the primary reason. In addition, affordability was a more significant factor for Black and Hispanic households.

Table 7. Main Reason Provided by Internet Non-Users, by Metropolitan Status, 2009

	All	Urban	Rural
Distribution of main reason by households (%)			
Don't need it - not interested	47.2	46.0	51.5
Too expensive	18.6	19.5	15.6
Can use it somewhere else	1.4	1.6	0.7
Not available in area	0.7	0.5	1.1
No computer or computer inadequate	22.3	22.1	23.0
Privacy and security	0.3	0.3	0.3
Concern for children's access	0.1	0.1	0.1
Lack of confidence or skill	4.3	4.5	3.7
Other reasons	5.1	5.4	4.1
Sample size	12,467	8,902	3,469
Estimated number of households	27,821,275	21,585,515	6,042,974

Source: U.S. Census Bureau, Current Population Survey (CPS) and CPS School Enrollment and Internet Use Supplement, October 2009, and ESA calculations..

Note: The sample size for urban and rural households does not add up to the total sample size because metropolitan-nonmetropolitan status is not identified for a small number of households. The share of households not using the Internet in 2009 was 23% for all households, 22% for urban households, and 32% for rural households.

Table 8. Main Reason Provided by Internet Non-Users, by Income, 2009

	All	Income Less than $25,000	Income $25,000- $50,000	Income $50,000- $75,000	Income $75,000- $100,000
Distribution of main reason by households (%)					
Don't need it - not interested	47.2	42.0	48.3	47.7	54.0
Too expensive	18.6	22.2	17.9	14.8	7.4
Can use it somewhere else	1.4	1.0	2.2	2.5	4.4
Not available in area	0.7	0.5	0.8	1.4	2.8
No computer or computer inadequate	22.3	24.4	21.0	22.6	19.4
Privacy and security	0.3	0.2	0.4	1.1	0.8
Concern for children's access	0.1	0	0.2	0	0.7
Lack of confidence or skill	4.3	4.8	3.6	4.5	4.3
Other reasons	5.1	5.0	5.6	5.3	6.2
Sample size	12,467	5,550	2,619	643	183
Estimated number of households	27,821,275	12,410,004	5,738,875	1,416,225	399,577

Source: U.S. Census Bureau, Current Population Survey (CPS) and CPS School Enrollment and Internet Use Supplement, October 2009, and ESA calculations.

Note: The share of households not using the Internet in 2009 was 23% for all households, 47% for households with incomes less than $25,000, 22% for households with incomes between $25,000 and $50,000, 8% for households with incomes between $50,000 and $75,000, and 4% for households with incomes between $75,000 and $100,000. The highest income category (income exceeding $100,000) was omitted because only a small minority, 2%, of households in this category did not use the Internet.

Table 9. Main Reason Provided by Internet Non-Users, by Race and Ethinicity, 2009

	All	Non-Hispanic White	Non-Hispanic Black	Hispanic
Distnbution of main reason by households (%)				
Don't need it - not interested	47.2	53.0	40.7	35.0
Too expensive	18.6	14.1	23.5	29.4
Can use it somewhere else	1.4	1.1	1.8	1.9
Not available in area	0.7	0.7	0.4	0.5
No computer or computer inadequate	22.3	21.1	23.6	26.0
Privacy and security	0.3	0.3	0.3	0.1
Concern for children's access	0.1	0.1	0.1	0.2
Lack of confidence or skill	4.3	4.3	3.9	3.2
Other reasons	5.1	5.3	5.7	3.7
Sample size	12,467	8,259	1,778	1,824
Estimated number of households	27,821,275	16,862,626	4,774,134	4,978,057

Source: U.S. Census Bureau, Current Population Survey (CPS) and CPS School Enrollment and Internet Use Supplement, October 2009, and ESA calculations..

Note: The share of households not using the Internet in 2009 was 23% for all households, 20% for White households, 32% for Black households, and 36% for Hispanic households.

Section 5.2. Among Households Using the Internet Outside of Home

Figure 1 showed that 8% of American households in 2009 did not access the Internet from home but used it elsewhere. This section will analyze the reasons given by these households for not subscribing to home broadband Internet services. This group represents 22% of those that do not use broadband Internet at home.

Unlike the primary reason provided by Internet non-users (lack of need or interest), the most commonly cited reason by households that did not use the Internet specifically at home was related to affordability or cost. This is not surprising since these persons used the Internet, but not at home, demonstrating their perceived interest and need for high-speed Internet. Figure 8 shows that 40% of households that used the Internet at a place other than home cited "too expensive" as the main impediment to using broadband Internet at home. Another 17% cited lack of need or interest, 17% cited lack of an adequate computer, and 15% cited the ability to use it somewhere else.

Table 10 shows the results for the overall sample as well as for urban and rural households. The two right columns of Table 10 show that urban and rural households rank these factors in a similar manner. Lack of availability, cited by 5% of rural households and 2% of urban households, was not a major deterrent to home broadband Internet use. As we will see in the next section, this is not the case for households using a dial-up Internet service.

Table 10. Main Reason Provided by Households Using the Internet Outside of Home, by Metropolitan Status, 2009

	All	Urban	Rural
Distribution of main reason by households (%)			
Don't need it — not interested	16.7	16.5	17.9
Too expensive	39.7	40.6	36.5
Can use it somewhere else	14.8	14.7	14.9
Not available in area	2.7	2.1	5.0
No computer or computer inadequate	16.9	16.9	16.5
Privacy and security	0.3	0.3	0.1
Concern for children's access	0.4	0.2	1.1
Lack of confidence or skill	0.4	0.5	0.3
Other reasons	8.2	8.3	7.8
Sample size	4,295	3,158	1,097
Estimated number of households	9,522,716	7,582,552	1,851,736

Source: U.S. Census Bureau, Current Population Survey (CPS) and CPS School Enrollment and Internet Use Supplement, October 2009, and ESA calculations..

Note: The sample size for urban and rural households does not add up to the total sample size because metropolitan-nonmetropolitan status is not identified for a small number of households. The share of households using the Internet at a location other than home in 2009 was 8% for all households, 8% for urban households, and 10% for rural households.

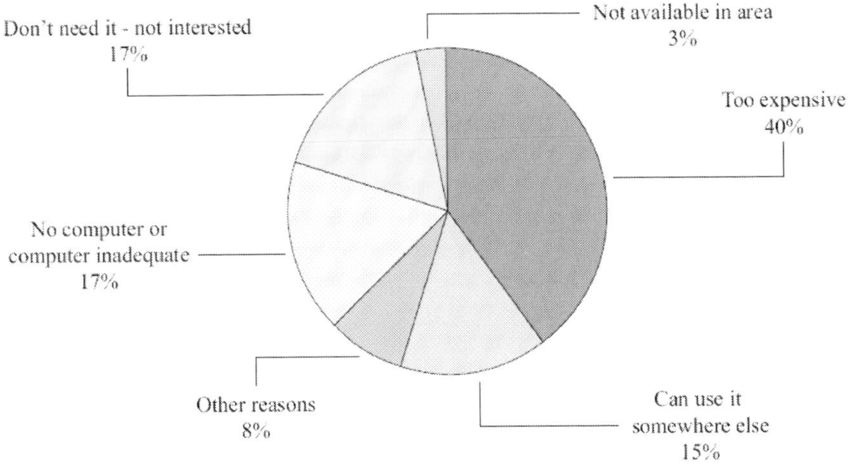

Source: U.S. Census Bureau, Current Population Survey (CPS) and CPS School Enrollment and Internet Use Supplement, October 2009, and ESA calculations.
Note: The figure does not include the categories that were reported as the main reason by less than 1% of non-users. See column 1 of Table 10 for the full set of results.

Figure 8. Main Reason Provided by Households Using the Internet Outside of Home, 2009.

Table 11 identifies the major reasons for not using broadband Internet at home by income categories. Affordability was the primary factor for households in the two lower income categories whereas other factors, like lack of demand, availability somewhere else, and lack of an adequate computer, were at least as important as affordability for households with incomes between $75,000 and $100,000. This suggests that affordability was among the top issues for Internet users that did not purchase home broadband Internet services, and this is true across a broad range of incomes.

Table 12 shows the primary reasons for not using broadband Internet at home by race and ethnicity. The most important reason was once again related to expense. However, a larger share of Black and Hispanic households (46% and 47%, respectively) than White households (35%) cited expense as the primary deterrent.

Table 11. Main Reason Provided by Households Using the Internet Outside of Home, by Income, 2009

	All	Income Less than $25,000	Income $25,000-$50,000	Income $50,000-$75,000	Income $75,000-$100,000
Distribution of main reason by households (%)					
Don't need it - not interested	16.7	8.5	17.8	27.3	26.2
Too expensive	39.7	53.6	35.3	26.6	17.6
Can use it somewhere else	14.8	10.3	15.9	17.8	21.6
Not available in area	2.7	1.0	3.4	2.6	8.6
No computer or computer inadequate	16.9	19.2	17.4	13.5	15.6
Privacy and security	0.3	0	0.5	0.3	0
Concern for children's access	0.4	0.4	0.4	0.6	0
Lack of confidence or skill	0.4	0.4	0.6	0.5	0.1
Other reasons	8.2	6.5	8.7	10.8	10.4
Sample size	4,295	1,452	1,229	535	193
Estimated number of households	9,522,716	3,282,895	2,675,640	1,119,851	397,588

Source: U.S. Census Bureau, Current Population Survey (CPS) and CPS School Enrollment and Internet Use Supplement, October 2009, and ESA calculations.

Note: The share of households using the Internet at a location other than home in 2009 was 8% for all households, 12% for households with incomes less than $25,000, 10% for households with incomes between $25,000 and $50,000, 6% for households with incomes between $50,000 and $75,000, and 4% for households with incomes between $75,000 and $100,000. The highest income category (income exceeding $100,000) was omitted because only a small minority, 2%, of households in this category used the Internet at a location other than home.

Table 12. Main Reason Provided by Households Using the Internet Outside of Home, by Race and Ethinicity, 2009

	All	Non-Hispanic White	Non-Hispanic Black	Hispanic
Distnbution of main reason by households (%)				
Don't need it - not interested	16.7	18.5	13.6	13.8
Too expensive	39.7	35.4	45.5	47.0
Can use it somewhere else	14.8	16.3	12.3	12.2
Not available in area	2.7	3.8	1.2	0.9
No computer or computer inadequate	16.9	15.5	20.6	17.4
Privacy and security	0.3	0.2	0.2	0.4
Concern for children's access	0.4	0.4	0.5	0.5
Lack of confidence or skill	0.4	0.4	0.3	0.6
Other reasons	8.2	9.3	5.7	7.2
Sample size	4,295	2,714	747	579
Estimated number of households	9,522,716	5,500,807	2,026,284	1,533,738

Source: U.S. Census Bureau, Current Population Survey (CPS) and CPS School Enrollment and Internet Use Supplement, October 2009, and ESA calculations.

Note: The share of households using the Internet at a location other than home in 2009 was 8.0% for all households, 7% for White households, 14% for Black households, and 11% for Hispanic households.

Section 5.3. Among Households with Dial-up Internet Access

Figure 1 showed that 5% of American households in 2009 used a dial-up telephone service to access the Internet from home. This group comprises the underlying sample for Figure 9 and Tables 13 through 15 (representing 13% of those that do not access broadband Internet at home), and their responses for why they use a dial-up service, as opposed to a broadband Internet connection, are tabulated in these tables.

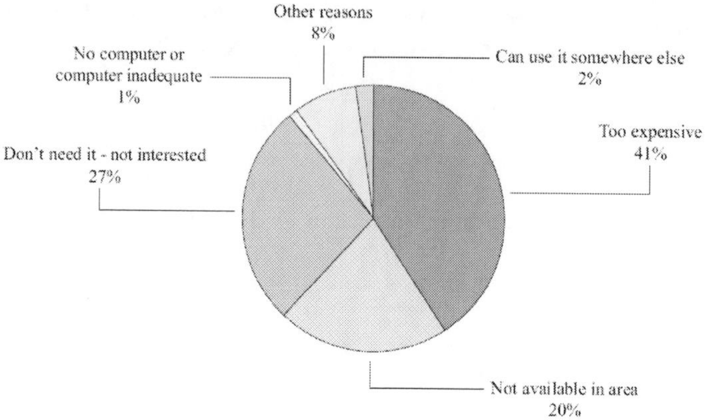

Source: U.S. Census Bureau, Current Population Survey (CPS) and CPS School Enrollment and Internet Use Supplement, October 2009, and ESA calculations.

Note: The figure does not include the categories that were reported as the main reason by less than 1% of non-users. See column 1 of Table 13 for the full set of results.

Figure 9. Main Reason Provided by Households with Dial-up Internet Access, 2009.

Table 13. Main Reason Provided by Households with Dial-up Internet Access, by Metropoliten Status, 2009

	All	Urban	Rural
Distribution of main reason by households (%)			
Don't need it — not interested	27.3	29.9	18.8
Too expensive	41.3	42.7	37.1
Can use it somewhere else	1.6	1.8	1.0
Not available in area	19.9	14.7	36.1
No computer or computer inadequate	1.1	1.4	0.2
Privacy and security	0.3	0.2	0.4
Concern for children's access	0.1	0.2	0
Lack of confidence or skill	0.8	0.9	0.4
Other reasons	7.6	8.2	6.1
Sample size	2,639	1817	799
Estimated number of households	5,648,799	4,263,953	1,341,764

Source: U.S. Census Bureau, Current Population Survey (CPS) and CPS School Enrollment and Internet Use Supplement, October 2009, and ESA calculations.

Note: The sample size for urban and rural households does not add up to the total sample size because metropolitan-nonmetropolitan status is not identified for a small number of households. The share of households using a dial-up Internet access in 2009 was 5% for all households, 4% for urban households, and 7% for rural households.

Figure 9 shows that the most commonly cited reason among dial-up users for not subscribing to broadband Internet at home in 2009 was "too expensive" (41%), followed by "don't need it - not interested" (27%), and lack of availability (20%).

Table 13 shows the results for the overall sample as well as for urban and rural households. The two right columns of Table 13 show that urban and rural dial-up users weighted these reasons differently, however. Lack of availability was as important as affordability for rural dial-up users. Slightly more than one-third (36%) of rural households with dial-up said that lack of broadband availability was their primary reason for not using broadband Internet services, compared to a much smaller share (15%) of their urban counterparts. This implies that lack of availability (or at least a perceived lack of availability) is a more significant deterrent in rural areas than urban areas.

Table 14 identifies the reasons for not purchasing broadband Internet service by income levels. The most important reason, once again, was related to affordability for all but the group of households with incomes between $75,000 and $100,000 (lack of demand, expense, and lack of availability were reported by approximately the same share of households in this group). This means that affordability is a major concern for households across a broad range of incomes. A perceived lack of demand was equally important across the income groups, whereas lack of availability was more important for higher income households.

Table 14. Main Reason Provided by Households with Dial-up Internet Access, by Income, 2009

	All	Income Less than $25,000	Income $25,000-$50,000	Income $50,000-$75,000	Income $75,000-$100,000
Distribution of main reason by households (%)					
Don't need it - not interested	27.3	25.9	25.4	27.1	25.5
Too expensive	41.3	49.5	44.0	36.7	30.7
Can use it somewhere else	1.6	1.3	1.0	1.9	3.6
Not available in area	19.9	14.7	18.9	25.3	25.9
No computer or computer inadequate	1.1	1.4	0.8	1.9	0.7
Privacy and security	0.3	0.2	0.2	0.3	0.4
Concern for children's access	0.1	0	0.3	0	0.8

Table 14. (Continued)

	All	Income Less than $25,000	Income $25,000-$50,000	Income $50,000-$75,000	Income $75,000-$100,000
Lack of confidence or skill	0.8	0.6	0.9	1.0	0
Other reasons	7.6	6.5	8.6	5.7	12.4
Sample size	2,639	532	783	472	192
Estimated number of households	5,648,799	1,155,790	1,608,936	993,354	413,788

Source: U.S. Census Bureau, Current Population Survey (CPS) and CPS School Enrollment and Internet Use Supplement, October 2009, and ESA calculations.

Note: The share of households using dial-up Internet access in 2009 was 5% for all households, 4% for households with incomes less than $25,000, 6% for households with incomes between $25,000 and $50,000, 6% for households with incomes between $50,000 and $75,000, and 4% for households with incomes between $75,000 and $100,000. The highest income category (income exceeding $100,000) was omitted because only a small minority, 2%, of households in this category had a dial-up Internet access.

Table 15. Main Reason Provided by Households with Dial-up Internet Access_ by Race and Ethnicity, 2009

	All	Non-Hispanic White	Non-Hispanic Black	Hispanic
Distnbution of main reason by households (%)				
Don't need it - not interested	27.3	27.5	28.8	22.0
Too expensive	41.3	37.8	50.0	55.5
Can use it somewhere else	1.6	1.3	1.3	4.7
Not available in area	19.9	23.3	13.5	6.2
No computer or computer inadequate	1.1	1.2	0.5	1.1
Privacy and security	0.2	0.3	0.5	0
Concern for children's access	0.1	0.2	0	0
Lack of confidence or skill	0.8	0.7	1.1	0.7
Other reasons	7.6	7.8	4.3	9.8
Sample size	2,639	2,033	265	229
Estimated number of households	5,648,799	4,147,090	653,916	600,717

Source: U.S. Census Bureau, Current Population Survey (CPS) and CPS School Enrollment and Internet Use Supplement, October 2009, and ESA calculations.

Note: The share of households using dial-up Internet access in 2009 was 5% for all households, 5% for White households, 4% for Black households, and 4% for Hispanic households.

Table 15 shows similar trends, looking at separate race and ethnic groups. The most important reason for not using broadband Internet was related to expense, although a larger share of Hispanic and Black households (56% and 50%, respectively) than White households (38%) reported "too expensive" as their primary impediment. Lack of availability was more important for White and Black households than Hispanic households. This means a much smaller share of Hispanic households felt impacted by a perceived lack of availability than White and Black households.

Section 6. Demographic and Geographic Characteristics of Broadband versus Dial-up Internet Users

This section compares household-level attributes among broadband-using and dial-up-using households. Households that utilized a dial-up Internet service in 2009 accounted for a small minority (7%) of households using the Internet at home, whereas broadband users accounted for the vast majority (93%). The share of dial-up Internet users is shrinking, and the decline in dial-up users has been more than offset by the expansion in broadband-Internet-using households. It is still useful to analyze how dial-up-using households differ from their broadband-using counterparts because such a comparison can help to explain lags in technology adoption.

Tables 16 and 17 show the distributions of income, education, and a number of other demographic and geographic characteristics by dial-up-using and broadband-using households. According to these tables, dial-up users in 2009 were, on average, older, had lower levels of family income and education, and were more likely to reside in rural areas.

Dial-up Internet users were, on average, less affluent than broadband Internet users. Compared to households that subscribed to broadband Internet service, a larger share of dial-up users had family incomes less than $25,000 and a smaller share had incomes exceeding $100,000 (Table 16). Dial-up Internet users also obtained less education than broadband Internet users. Compared to broadband Internet users, a lower fraction of dial-up users had a college degree and a higher share had less than a high school degree.

Dial-up Internet users were older with an average age of 54, compared to 47 among broadband Internet users. A larger share of dial-up-using households was headed by someone with a disability, 16%, compared to half of that (8%) among their broadband-Internet-using counterparts.

Table 16. Household Demographic Attributes: Dial-up versus Broadband Internet Users, 2009

	Distribution by Households using Dial-up Internet (%)	**Distribution by Households using Broadband Internet (%)**
Percent of All Households	4.7	63.5
Percent of Households connecting to the Internet from home	6.9	92.5
Family Income		
Less than $25,000	20.5	12.5
$25,000-$50,000	28.5	20.9
$50,000-$75,000	17.6	18.4
$75,000-$100,000	7.3	11.7
More than $100,000	4.7	18.6
Education		
Less than High School Degree	10.8	5.6
High School Degree	36.4	23.5
Some College	29.7	30.9
College Degree or more	23.1	40.0
Race and Ethnicity		
White, Non-Hispanic	73.4	75.2
Black, Non-Hispanic	11.6	9.7
Asian, Non-Hispanic	2.4	4.7
American Indian or Alaskan Native, Non-Hispanic	0.6	0.5
Hispanic	10.6	8.7
Foreign-Born Status		
Citizens (including foreign born)	93.9	94.8
Non-Citizen	6.1	5.2
Age (mean years)	**54.4**	**46.8**
Gender	**41.4**	**24.1**
Male	49.3	53.6
Female	50.7	46.4

	Distribution by Households using Dial-up Internet (%)	Distribution by Households using Broadband Internet (%)
Household Type		
Married-couple with children	20.0	27.4
Single parents (male)	2.1	2.3
Single parents (female)	7.1	7.3
Family without children	41.6	36.1
Non-family households	29.3	27.0
Disability Status		
Has a disability	15.7	8.4
No disability	84.2	91.1
Sample Size	2,639	34,633
Estimated Number of Households	5,648,799	75,776,370

Source: U.S. Census Bureau, Current Population Survey (CPS) and CPS School Enrollment and Internet Use Supplement, October 2009, and ESA calculations.

Note: The distributions across the income categories do not sum to 100% since income data are not reported by some households.

Table 17 looks at geographic attributes and shows that households using dial-up Internet services were more likely to be rural dwellers. Almost one in four (24%) dial-up Internet users lived in rural areas, compared to about one in eight broadband Internet users (13%). Compared to dial-up users, broadband Internet users were more likely to live in large urban areas with populations exceeding one million (57% of broadband users, compared to 43% of dial-up users, lived in urban areas with populations exceeding one million).

SECTION 7. DISABILITY AND BROADBAND INTERNET USE

This section analyzes broadband Internet adoption by people with disabilities. The disability community is a key population group that presents special challenges for adoption of broadband Internet and other modes of communications. "Disability" is defined by the Americans with Disabilities Act as "a physical or mental impairment that substantially limits a major life activity" (U.S. Department of Justice, 2010).[8] There were 36.1 million people with disabilities in 2008, or about 12.1% of the population (U.S. Census Bureau, 2010(b)). More than half have severe disabilities (Lyle, 2010).

The analysis presented in Section 4 of this report showed that people with disabilities were less likely than people with no disabilities to use broadband Internet at home, and that differences in socioeconomic characteristics explain the majority of the adoption gap associated with disability. This section first looks at the profile of people with disabilities and their Internet and broadband Internet usage patterns. We then study their main reasons for non-use. In order to be consistent with the rest of the report, this section looks at broadband Internet (or Internet) adoption at home and its association with disability status of the head of household. Note that people with disabilities in this section refer to household heads with disabilities.[9]

Table 17. Household Geographic Attributes: Dial-up versus Broadband Internet Users, 2009

	Distribution by Households using Dial-up Internet (%)	Distribution by Households using Broadband Internet (%)
Region		
Northeast	17.1	19.2
Midwest	24.6	22.0
South	35.8	34.9
West	22.5	23.9
Urban-Rural Status		
Urban	75.5	86.6
Rural	23.8	12.8
Metropolitan Area (CBSA) Size		
Under 1,000,000	28.0	26.9
1,000,000-2,499,999	14.6	18.1
2,500,000-4,999,999	12.4	19.1
5,000,000 or more	16.2	19.5
Sample Size	2,639	34,633
Estimated Number of Households	5,648,799	75,776,370

Source: U.S. Census Bureau, Current Population Survey (CPS) and CPS School Enrollment and Internet Use Supplement, October 2009, and ESA calculations.

Section 7.1. Profile of People with Disabilities

Table 18 presents data on income, education, age, and geographic location for the entire population, and separately by disability status of the head of the household. Fourteen percent of household heads, representing almost 17 million people, had a disability in 2009.[10] People with disabilities, on average, were older with an average age of 63, compared to 48 among householders with no disability. People with disabilities also had lower levels of household income and obtained less education. Almost half of all householders with disabilities (45%) had family incomes less than $25,000, compared to a fifth (19%) of the population with no disability. One in four people with disabilities (25%) did not have a high school degree, compared to one in ten among people with no disability (10%). People with disabilities were also more likely to live in rural areas-22% of households where the householder had a disability lived in rural areas, compared to 15% of those with no disability.

Table 18. Household Demographic and Geographic Characteristics by Disability Status, 2009

	All Households	Distribution by Households/Householder where the householder has a disability (%)	Distribution by Households/Householder where the Householder has no disability (%)
Percent of All Households	100	14.1	85.6
Family Income			
Less than $25,000	22.1	44.9	18.5
$25,000-$50,000	21.8	19.3	22.1
$50,000-$75,000	14.7	8.6	15.7
$75,000-$100,000	8.5	3.4	9.3
More than $100,000	12.5	3.6	14.0
Education			
Less than High School Degree	12.3	24.7	10.3
High School Degree	29.3	34.4	28.5
Some College	28.3	25.5	28.6
College Degree or more	30.1	15.4	32.5
Age (mean years)	49.7	63.0	47.6

Table 18. (Continued)

	All Households	Distribution by Households/Householder where the householder has a disability (%)	Distribution by Households/Householder where the Householder has no disability (%)
Geographic Location			
Urban	83.4	77.2	84.4
Rural	15.9	22.1	14.9
Sample Size	54,280	7,935	46,140
Estimated Number of Households	119,267,400	16,768,677	102,062,153

Source: U.S. Census Bureau, Current Population Survey (CPS) and CPS School Enrollment and Internet Use Supplement, October 2009, and ESA calculations.

Note: The distributions across the income categories do not sum to 100% since income data are not reported by some households.

Section 7.2. Internet Use by People with Disabilities

Given that people with disabilities have lower levels of income and education, are older, and are more likely to reside in rural areas, it is not surprising that they display lower broadband and Internet usage rates. Table 19 shows Internet and broadband Internet usage rates for the entire population, and separately by disability status of the household head. Only half of all households (51%) headed by someone with a disability had an Internet user, compared to the majority of households (81%) headed by someone without a disability. Only four out of ten households (43%) headed by someone with a disability subscribed to Internet services at home, compared to seven out of ten households (73%) where the householder has no disability. Broadband Internet subscription at home showed a similar pattern (38% compared to 68%).

Section 7.3. Main Reason for Non-Adoption by People with Disabilities

Section 4 of this report showed that socio-economic and geographic characteristics explained most of the adoption gap associated with disability.

This report now looks at the main reason for non- adoption for people with disabilities. Tables 20, 21 and 22 analyze the main reasons for not having broadband Internet access at home for three types of non-users — households that do not use the Internet at any location, households that do not use the Internet specifically at home, and households that use dial-up Internet at home. These tables show that the primary reasons for non-adoption are largely similar for people with and without disabilities.

Table 20 shows that the main reason provided by Internet non-users was lack of demand, regardless of disability status. Compared to people with no disabilities, a smaller share of people with disabilities gave cost as the major reason (14% compared to 21%) and a slightly larger share of people with disabilities gave lack of confidence or skill as the major reason (6% compared to 4%).

Unlike Internet non-users, affordability was the biggest impediment to home broadband Internet access for households that did not use the Internet specifically at home (Table 21). A larger share of households headed by someone with a disability provided affordability as the principal reason (47% compared to 39%).

Affordability was among the top concerns for households with dial-up Internet access and was selected by the same share of such households, regardless of disability status.

Table 19. Average Internet Use by Disability Status, 2009

	All Households	Households where the Householder has a disability	Households where the Householder has no disability
Percent of All Households	100	14.1	85.6
Internet Use (%)			
At any location	76.7	50.6	80.9
At home	68.7	43.4	72.7
Use broadband Internet at home	63.5	37.8	67.6
Use dialup Internet at home	4.7	5.3	4.7
Sample Size	54,280	7,935	46,140
Estimated Number of Households	119,267,400	16,768,677	102,062,153

Source: U.S. Census Bureau, Current Population Survey (CPS) and CPS School Enrollment and Internet Use Supplement, October 2009, and ESA calculations.

Table 20. Main Reason Provided by Internet Non-Users, by Disability Status, 2009

	All	With Disability	No Disability
Distribution of main reason by households (%)			
Don't need it - not interested	47.2	48.8	46.6
Too expensive	18.6	13.6	20.7
Can use it somewhere else	1.4	0.5	1.8
Not available in area	0.7	0.4	0.8
No computer or computer inadequate	22.3	24.0	21.6
Privacy and security	0.3	0.4	0.3
Concern for children's access	0.1	0.0	0.1
Lack of confidence or skill	4.3	6.0	3.6
Other reasons	5.1	6.4	4.6
Sample size	12,467	3,912	8,548
Estimated number of households	27,821,275	8,287,550	19,520,186

Source: U.S. Census Bureau, Current Population Survey (CPS) and CPS School Enrollment and Internet Use Supplement, October 2009, and ESA calculations.

Note: The share of households not using the Internet in 2009 was 23% for all households, 49% for households headed by someone with a disability, and 19% for households headed by someone with no disability.

Table 21. Main Reason Provided by Households Using the Internet Outside of Home, by Disability Status, 2009

	All	With Disability	No Disability
Distribution of main reason by households (%)			
Don't need it - not interested	16.5	14.6	17.0
Too expensive	39.7	47.4	38.6
Can use it somewhere else	14.8	10.3	15.4
Not available in area	2.7	2.7	2.7
No computer or computer inadequate	16.9	17.5	16.8
Privacy and security	0.3	0.5	0.2
Concern for children's access	0.4	0.4	0.4
Lack of confidence or skill	0.4	0.3	0.5
Other reasons	8.2	6.4	8.4
Sample size	4,295	552	3,740
Estimated number of households	9,522,716	1,205,891	8,309,436

Source: U.S. Census Bureau, Current Population Survey (CPS) and CPS School Enrollment and Internet Use Supplement, October 2009, and ESA calculations.

Note: The share of households using the Internet at a location other than home in 2009 was 8% for all households, 7% for households headed by someone with a disability, and 8% for households headed by someone with no disability.

Table 22. Main Reason Provided by Households with Dial-up Internet Access, by Disability Status, 2009

	All	With Disability	No Disability
Distribution of main reason by households (%)			
Don't need it — not interested	27.3	27.4	27.3
Too expensive	41.3	39.6	41.6
Can use it somewhere else	1.6	1.5	1.7
Not available in area	19.9	21.0	19.7
No computer or computer inadequate	1.1	1.5	1.0
Privacy and security	0.3	0	0.3
Concern for children's access	0.1	0	0.2
Lack of confidence or skill	0.8	1.2	0.7
Other reasons	7.6	7.8	7.6
Sample size	2,639	425	2,212
Estimated number of households	5,648,799	888,070	4,754,852

Source: U.S. Census Bureau, Current Population Survey (CPS) and CPS School Enrollment and Internet Use Supplement, October 2009, and ESA calculations.

Note: The share of households using dial-up Internet access in 2009 was 5% for all households, 5% for households headed by someone with a disability, and 5% for households headed by someone with no disability.

SECTION 8. LONG-TERM COMPARISONS: 2001 VERSUS 2009

The last section of this report will analyze the change in home broadband Internet adoption over this decade by comparing the most recent data from 2009 with that collected in 2001.[11] Sections 8.1, 8.2, and 8.3 analyze the growth in broadband Internet adoption over this decade by household demographic and geographic characteristics. Section 8.4 compares the marginal effects over time of selected demographic and geographic characteristics on the likelihood of home Internet use (Section 8.4.1) and broadband Internet use (Section 8.4.2).

Section 8.1. Broadband Internet Use by Demographic Characteristics: 2001 versus 2009

Table 23 presents data on average broadband Internet usage rates by demographic characteristics in 2001 and 2009. Table 23 shows that the share of households with broadband Internet service has risen sevenfold between 2001 and 2009, from 9% to 64% of households using broadband Internet services at home. Some of the groups which began with much lower adoption rates in 2001 have since exhibited significant gains. These impressive gains, however, have not eliminated the gaps within demographic groups defined by income, education, race, ethnicity, and age. For example, households in the lowest income group, with annual incomes less than $25,000, exhibited a twelvefold rise in broadband Internet adoption, from 3% to 36%, while households with incomes exceeding $75,000 saw more than a fourfold rise, from 21% to 92%. Despite the faster rise in the share of lower-income households with broadband Internet, a sizeable gap in average connectivity between these two groups, 36% compared to 92%, still persisted in 2009.

Table 23. Household Broadband Internet Use by Demographic Characteristics, 2001 and 2009

	Percent of households connecting to the Internet at home using broadband	
	2001	**2009**
All	9.2	63.5
Family Income		
Less than $25,000	3.1	35.8
$25,000-$50,000	7.3	61.0
$50,000-$75,000	12.2	79.3
More than $75,000	20.8	91.5
Education		
Less than High School Degree	2.4	28.8
High School Degree	5.7	50.9
Some College	10.3	69.5
College Degree or more	16.3	84.5
Race and Ethnicity*		
White, Non-Hispanic	10.2	68.0

	Percent of households connecting to the Internet at home using broadband	
	2001	2009
Black, Non-Hispanic	4.7	49.4
Asian, Non-Hispanic	15.1	77.3
American Indian or Alaskan Native, Non-Hispanic	7.6	48.3
Hispanic	5.3	47.9
Foreign-Born Status		
Citizens (including foreign born)	9.3	64.4
Non-citizen	7.7	51.0
Age		
16 to 44 years	11.3	71.2
45 to 64 years	10.1	68.2
65 years and over	3.1	39.9
Gender		
Male	10.7	66.7
Female	7.6	60.2
Household Type		
Married-couple with Children	13.9	79.8
Single parents (male)	7.2	60.1
Single parents (female)	5.6	56.9
Family without children	9.1	67.7
Non-family households	7.0	50.8
Sample Size	56,573	54,280
Estimated Number of Households	107,064,178	119,267,400

Source: U.S. Census Bureau, Current Population Survey (CPS) and CPS School Enrollment and Internet Use Supplement, October 2009, CPS and CPS Computer and Internet Use Supplement, September 2001, and ESA calculations.

Note: *Contrary to 2009, the race and ethnicity indicators in 2001 do not separately identify multi-race categories. As a result, the race and ethnicity categories are not strictly comparable between 2001 and 2009.

A similar pattern holds for other demographic groups. For instance, households where the householder had at least a college degree were much more likely in 2009 to have broadband Internet than their counterparts with a high school degree or less, even though households headed by someone with a high school degree or less experienced faster growth in broadband Internet use. Similarly, Hispanic households and non-Hispanic Black households had broadband Internet adoption rates which were half of non-Hispanic White adoption rates in 2001. Both Hispanic households and Black non-Hispanic households exhibited large gains in connectivity, but substantial gaps in adoption persisted across the race and ethnic groups in 2009. In 2001, the broadband Internet adoption rate for seniors (3%) was about one-fourth of the average rate for those between 16 and 44 years of age (11%). In 2009, the average senior adoption rate was slightly more than half of that for the 16-44 year old group (40% and 71%, respectively).

All groups have seen impressive growth during this decade in the use of home broadband Internet services, which has resulted in a "catching-up" between low adopters and high adopters over time. However, significant adoption gaps persist along demographic and socio-economic dimensions.

Table 24. Household Broadband Internet Use by Geographic Location, 2001 & 2009

Percent of households connecting to the Internet at home using broadband		
	2001	**2009**
All Households	9.2	63.5
Region		
Northeast	11.3	67.0
Midwest	7.2	62.2
South	7.9	60.0
West	11.7	68.0
Urban-Rural Status		
Urban	10.5	65.9
Rural	3.8	51.0
Sample Size	56,573	54,280
Estimated Number of Households	107,064,178	119,267,400

Source: U.S. Census Bureau, Current Population Survey (CPS) and CPS School Enrollment and Internet Use Supplement, October 2009, CPS and CPS Computer and Internet Use Supplement, September 2001, and ESA calculations.

Note: The terms "urban" and "rural" refer to metropolitan and nonmetropolitan areas, respectively. Metropolitan area identifier in the 2001 CPS is based on Office of Management and Budget's 1990/1993 standards. According to defmitions adopted in 1990, the term "metropolitan area" collectively referred to metropolitan statistical areas (MSAs), consolidated metropolitan statistical areas (CMSAs), and primary metropolitan statistical areas (PMSAs). Metropolitan area identifier for 2009 is based on "core based statistical area" (CBSA) which refers collectively to metropolitan and micropolitan statistical areas.

Section 8.2.Broadband Internet Use by Geographic Region, and Urban and Rural Locations: 2001 versus 2009

Table 24 presents average broadband Internet usage rates by geographic location (region and urban-rural location) in 2001 and 2009. Table 24 portrays a picture very similar to the one seen in the previous section --- broadband Internet use among households living in geographic areas with historically low adoption rates rose faster than among their counterparts in high adoption areas but some differences, particularly between urban and rural locations, were still present in 2009. According to Table 24, average broadband Internet usage rates in the South and the Midwest were about two-thirds of the average rates in the Northeast and the West in 2001. Households in all four regions have seen gains in broadband Internet adoption, with those in the Midwest and the South experiencing faster gains and therefore narrowing the gap with their counterparts in the Northeast and the West. A similar pattern holds for urban and rural locations, although there remained a 15 percentage point urban-rural gap in home broadband Internet use in 2009.

Section 8.3. Broadband Internet Use by State: 2001 versus 2009

Table 25 ranks the states in descending order by their average broadband Internet adoption rates in 2001 and 2009. The analysis from Section 4 of this report noted that broadband Internet usage varied significantly across states, even after controlling for household demographic characteristics and urban and rural locations. It is therefore not surprising that average state-level broadband Internet adoption rates varied from 42% to 73% in 2009 and from 2% to 18% in 2001. Table 25 shows that the states in the Northeast and West regions generally exhibited higher broadband Internet access than those in the South and the Midwest. The state rankings were relatively constant between 2001 and 2009. For instance, out of the top 15 broadband-Internet-using states in 2001, 11 were still among the top 15 in 2009. The states in the bottom were somewhat unchanged also—out of the bottom 15 states in 2001, 9 states were still among the bottom 15 in 2009.

According to the findings in Section 4, demographic and geographic characteristics – primarily income, education, race, ethnicity, and the extent of urbanization - explain some but not all of the variation in broadband Internet adoption rates across states. This raises the following question: how much of the variation in adoption across states is driven by variation in broadband

Internet availability? The CPS does not provide information on availability of broadband Internet services at the household's location. In order to compare adoption with availability, we utilized state-level data from the Federal Communications Commission (FCC).

Table 25. Ranking of States by Average Home Broadband Internet Adoption, 2001 and 2009

2009 Ranking		2001 Ranking	
State	Percent of Households Using Broadband Internet at Home	State	Percent of Households Using Broadband Internet at Home
UT	73 (12)	HI	18
NH	73 (14)	MA	14
AK	73 (13)	NH	14
MA	73 (14)	CA	13
NJ	72 (12)	AK	13
WA	72 (11)	NJ	12
CT	71 (11)	AZ	12
OR	70 (9)	NY	12
HI	70 (18)	UT	12
MD	70 (9)	CT	11
RI	69 (11)	WA	11
CO	69 (10)	RI	11
NV	68 (7)	TN	10
CA	68 (13)	CO	10
ID	67 (6)	FL	10
AZ	67 (12)	KS	10
WI	67 (6)	NE	10
MN	67 (8)	ME	10
KS	67 (10)	OR	9
DE	67 (6)	TX	9
FL	67 (10)	MD	9
DC	66 (7)	MI	9
NY	66 (12)	SD	9
WY	66 (6)	IA	9
VA	65 (8)	SC	8
GA	64 (7)	PA	8

2009 Ranking		2001 Ranking	
State	Percent of Households Using Broadband Internet at Home	State	Percent of Households Using Broadband Internet at Home
NE	64 (10)	VT	8
IL	63 (6)	MN	8
ND	63 (5)	OK	8
MI	62 (9)	VA	8
IA	62 (9)	OH	8
PA	62 (8)	DC	7
OH	61 (8)	MO	7
ME	61 (10)	NV	7
VT	61 (8)	GA	7
TX	60 (9)	NC	7
SD	60 (9)	ID	6
NC	59 (7)	IL	6
MT	58 (3)	DE	6
MO	57 (7)	WI	6
LA	57 (5)	WY	6
IN	56 (4)	MS	6
OK	56 (8)	LA	5
TN	55 (10)	AL	5
NM	55 (2)	ND	5
KY	54 (3)	WV	4
SC	53 (8)	AR	4
WV	52 (4)	IN	4
AR	51 (4)	MT	3
AL	48 (5)	KY	3
MS	42 (6)	NM	2

Source: U.S. Census Bureau, Current Population Survey (CPS) and CPS School Enrollment and Internet Use Supplement, October 2009, CPS and CPS Computer and Internet Use Supplement, September 2001, and ESA calculations.

Note: The numbers in parentheses next to the 2009 data refer to the 2001 adoption rate for the state. Because of sampling variability, average adoption rates for two states may not be different from one another in a statistically significant way. Tables A7 and A8 in Appendix Section A6 provide the standard error and 90% confidence interval for each state.

The FCC collects data on high-speed Internet access services — both on adoption (number of high-speed Internet connections by type and speed of technology and by state) and availability (number of high-speed Internet service providers, by census tract level and state) (FCC, 2010). In order to compare the adoption measure used in this report with that from the FCC, we compared average state-level broadband Internet adoption from the CPS with FCC's data on the number of high-speed residential connections by state. We found that the two independent measures of adoption were highly positively correlated, validating the reliability of the home broadband Internet adoption measure from the CPS. In order to compare adoption with availability, we utilized three measures of availability at the state level from the FCC: total number of high-speed Internet service providers by state, percent of residences where the local telephone service providers provide DSL, and percent of residences where cable TV service providers provide cable modem Internet services. We found only a weak positive association between our estimated broadband Internet adoption and the data on availability. This weak correlation between adoption and availability at the state level is not surprising. Section 5 on main reasons for non-adoption showed that lack of availability primarily impacted rural households that did not subscribe to a highspeed broadband Internet service at home and used a slower dial-up service instead. This suggests that one is likely to find a stronger association between adoption and availability only when using more granular geographic data. State-level data are aggregated across urban and rural areas and are therefore likely to show little correlation between adoption and availability.

Section 8.4. Marginal Effects of Demographic and Geographic Characteristics on Adoption over Time

Given the results of this section, one must ask how the adoption gaps, after accounting for differences in demographic and geographic characteristics, have changed over the decade. In other words, how robust are the marginal effects of household characteristics on adoption? Section 4.2 of this report utilized a regression analysis framework that estimated the impact of multiple factors together on the probability that a household adopted broadband Internet services at home. The analysis allowed us to isolate the effect of any one factor, holding everything else constant. We referred to these results as the marginal effects of selected demographic and geographic characteristics on household broadband Internet use. The factors that we controlled for included family income, education, age, race, ethnicity, foreign-born status, household size (total number of persons in household), disability status, and geographic location (urban-rural location and state).

In this section we will apply a similar regression analysis framework to estimate the marginal effects of demographic and geographic factors on home Internet use in 2001 and 2009, and the marginal effects of demographic and geographic factors on home broadband Internet use in 2007 and 2009. Why do we focus on home Internet use generally in lieu of home broadband Internet use for the comparative analysis for 2001 and 2009? Broadband Internet was a relatively new technology in the early 2000s, which is why both availability and adoption levels at that time were low but both grew significantly over the decade. In order to compare the impacts of household characteristics on adoption between two time periods, it is useful to use a metric or indicator of adoption that is consistent over time. This is why we will first focus on Internet use at home to compare the adoption gaps or marginal effects between 2001 and 2009 (Section 8.4.1). In order to compare the adoption gaps or marginal effects for home broadband Internet use, we will use data for 2007 and 2009 (Section 8.4.2).

The full set of regression results from these analyses is presented in Sections A7 and A8 of the Appendix. The factors that we control for in these analyses include household income, education, age, race, ethnicity, foreign-born status, household size (total number of persons in household), and geographic location (urban-rural location and state).[12]

Section 8.4.1. Marginal Effects of Demographic and Geographic Characteristics on the Likelihood of Home Internet Use, 2001 versus 2009

Between 2001 and 2009, Internet use at home rose by one-third, from 51% to 69% of American households connecting to the Internet from home. Table 26 presents the estimated marginal effects of selected socioeconomic (income, education, race, ethnicity), and geographic (urban-rural status) factors on the probability of Internet use at home. These marginal effects changed only slightly between 2001 and 2009. For example, after controlling for various non-income attributes, the gap between households with incomes exceeding $75,000 and households with incomes less than $25,000 declined slightly, from 37 to 32 percentage points. Compared to households where the householder has less than a high school degree, the relative gains associated with a high school degree and some college have risen but those associated with a college degree or more have not changed.

Table 26. Marginal Effects of Demographic and Geographic Characteristics on the Likelihood that a Household Uses Internet at Home, 2001 and 2009

Household Characteristics	Adoption Gap in 2001 (percentage point)	Adoption Gap in 2009 (percentage point)
Household Income		
Gap between households with incomes $25,000-$50,000 and households with incomes less than $25,000	16	18
Gap between households with incomes $50,000-$75,000 and households with incomes less than $25,000	30	30
Gap between households with incomes exceeding $75,000 and households with incomes less than $25,000	37	32
Education		
Gap between those with a high school degree and those with less than high school degree	10	13
Gap between those with some college and those with less than high school degree	21	25
Gap between those with a college degree or more and those with less than high school degree	29	30
Race and Ethnicity		
Gap between White, Non-Hispanic and Black, Non-Hispanic	16	10
Gap between White, Non-Hispanic and Hispanic	15	14
Gap between White, Non-Hispanic and Asian	-5	2
Urban-Rural		
Gap between urban and rural households	4	5
Sample Size	47,310	43,662
Estimated Number of Households	88,963,933	94,963,684

Source: U.S. Census Bureau, Current Population Survey (CPS) and CPS School Enrollment and Internet Use Supplement, October 2009, CPS and CPS Computer and Internet Use Supplement, September 2001 and ESA calculations.

Note: Sample includes all households with the head of household aged 16 or more, and non-missing data on income.

Table 26 also shows that the White-Black gap in home Internet use, after holding the other attributes constant, has fallen from 16 percentage points in 2001 to 10 percentage points in 2009, whereas the White-Hispanic gap remained largely unchanged. This suggests that Internet use in Black households, on average, has gotten closer to that in White households, after accounting for socioeconomic and geographic characteristics, although the same pattern of convergence does not hold for Hispanic households. The urban-rural gap in home Internet use stayed similar, implying that the increased likelihood of home Internet use by urban dwellers relative to their rural-area counterparts has not changed.

Section 8.4.2. Marginal Effects of Demographic and Geographic Characteristics on the Likelihood of Home Broadband Internet Use, 2007 versus 2009

Between 2007 and 2009, broadband Internet use among households rose by one-fourth, from 51% to 64% of American households using broadband Internet services. Table 27 compares the marginal effects of household characteristics on the likelihood of home broadband Internet use between 2007 and 2009. The full set of these regression results is presented in Section A8 of the Appendix. The marginal effects of household demographic and geographic factors on home broadband Internet adoption stayed largely unchanged between 2007 and 2009. For example, relative to households in the lowest income group, the likelihood of adoption rose slightly (from 13 to 17 percentage points) for households with incomes between $25,000 and $50,000, while the likelihood of adoption fell slightly (from 39 to 35 percentage points) for households in the highest income group.

Table 27. Marginal Effects of Demographic and Geographic Characteristics on the Likelihood that a Household Uses Broadband Internet at Home, 2007 and 2009

Household Characteristics	Adoption Gap in 2007 (percentage point)	Adoption Gap in 2009 (percentage point)
Household Income		
Gap between households with incomes $25,000-$50,000 and households with incomes less than $25,000	13	17
Gap between households with incomes $50,000-$75,000 and households with incomes less than $25,000	26	28
Gap between households with incomes $75,000-$100,000 and households with incomes less than $25,000	33	32

Table 27. (Continued)

Household Characteristics	Adoption Gap in 2007 (percentage point)	Adoption Gap in 2009 (percentage point)
Gap between households with incomes exceeding $100,000 and households with incomes less than $25,000	39	35
Education		
Gap between those with a high school degree and those with less than high school degree	8	11
Gap between those with some college and those with less than high school degree	21	23
Gap between those with a college degree or more and those with less than high school degree	30	29
Race and Ethnicity		
Gap between White, Non-Hispanic and Black, Non-Hispanic	11	10
Gap between White, Non-Hispanic and Hispanic	12	14
Gap between White, Non-Hispanic and Asian	-4	0
Urban-Rural		
Gap between urban and rural households	8	7
Sample Size	42,481	43,662
Estimated Number of Households	91,153,697	94,963,684

Source: U.S. Census Bureau, Current Population Survey (CPS) and CPS School Enrollment and Internet Use Supplement, October 2007 and October 2009, and ESA calculations.

Note: Sample includes all households with the head of household aged 16 or more, and non-missing data on income.

Table 27 also shows that the White-Black and the White-Hispanic gaps in home broadband Internet use, as well as urban-rural gaps, did not change considerably between 2007 and 2009.

SECTION 9. CONCLUSION

Household use of broadband Internet service has risen dramatically during this decade as the Internet has expanded to become an integral component of life for many American households. Nonetheless, not everyone uses the Internet or has access to it. This report attempts to analyze what factors are associated with home broadband Internet adoption, using data from a special 2009 supplement to the CPS that asked questions about broadband Internet use of more than 50,000 households.

The analysis determines that some significant adoption gaps exist today, particularly by income and education levels. For example, controlling for various non-income household attributes, the gap between households with incomes greater than $100,000 and those with incomes less than $25,000 totals 34 percentage points. Similarly, the controlled gap for those with at least college degrees versus those with no high school diplomas tallies 29 percentage points.

However, this report also shows that income and education levels, although strongly associated with broadband Internet use, are not the sole determinants of broadband Internet adoption by households. Even after accounting for differences in income and education (and a number of other key household attributes), there remain significant differences in adoption rates across race and ethnicity, and across urban and rural areas. For example, the unexplained gaps between Whites and Blacks (10 percentage points) and Whites and Hispanics (14 percentage points) remain at double digits in 2009. The adoption gap for rural versus urban is halved once controlled for household attributes, but still registers 7 percentage points.

This report also finds that lack of need or interest, lack of affordability, lack of an adequate computer, and lack of availability were all stated as main determinants of non-adoption of broadband Internet services. The significance of these factors, however, varied across non-users, with affordability and demand generally dominating. For instance, Internet non-users (representing almost two-thirds of non-users of broadband at home) reported lack of need or interest as the main reason for not subscribing to home broadband Internet services, whereas affordability was the most important deterrent for those who either used the Internet at a location other than home or used a dial-up Internet service at home. This suggests that those who used the Internet had a higher estimation of broadband Internet's value and need than those that did not. The significance of these factors differed somewhat by income, race, and ethnicity, but affordability appeared to be a major concern for households even at relatively high income levels.

This report also finds that broadband Internet use among households rose sevenfold, from 9% to 64%, between 2001 and 2009. Some of the groups that had lower than average adoption rates in 2001 have since exhibited impressive gains. Substantial adoption gaps still persist in broadband Internet access within demographic groups, as well as across states and between urban and rural areas.

Even though broadband Internet use has expanded significantly during this decade, not all groups are participating in the Internet revolution to the same extent. For a number of reasons, some groups lag behind in adopting this technology that has altered the social and economic landscape of the country. For example, broadband Internet adoption is particularly low among rural low-income Black and Hispanic households. This report enhances our understanding of the factors that are important drivers of — or impediments to — adoption. This information may contribute to the national efforts to ensure that everyone has an opportunity to be linked into the services and information available through the Internet.

APPENDIX

Section AI. Data

This report utilizes data from the Department of Commerce's U.S. Census Bureau, taken from the Census Bureau's October 2009 Current Population Survey (CPS) of 54,324 interviewed households. Data collection for the survey took place from October 18 through 26, 2009, and generated response rates of 92.1 percent for the basic CPS, with 93.8 percent of the CPS respondents answering the School Enrollment and Internet Use Supplement (i.e., 86.4 percent of the CPS sample answered the supplement).

The households surveyed were selected from the 2000 Decennial Census with coverage in all 50 states and the District of Columbia. The sample is continually updated to account for new residential construction. The Census divided the United States into 2,025 geographic areas, each typically comprised of a county or several contiguous counties. A total of 824 geographic areas were selected for the 2009 CPS survey. For each household, Census Bureau interviewers spoke to a person (called the "respondent") who was at least 15 years old and was considered knowledgeable about everyone in the household. The survey collected data both at the household level and at the individual level. For purposes of collecting data at an individual level, the respondent provided responses for himself or herself and proxy responses for all other members of that household age 3 and older. The survey, therefore, provided information on more than 129,000 individuals (age 3 and older).

Table Al. Individual Broadband Internet Use at the Individual Level, 2009

Individual Broadband Internet Use: Percent of persons connecting to the Internet at home using broadband	
All Persons*	59.1
Household Income	
Less than $25,000	31.9
$25,000-$50,000	52.4
$50,000-$75,000	70.0
$75,000-$100,000	78.7
More than $100,000	86.5
Education	
Less than High School Degree	37.3
High School Degree	48.1
Some College	70.6
College Degree or more	83.4
Race and Ethnicity	
White, Non-Hispanic	65.7
Black, Non-Hispanic	45.9
Asian, Non-Hispanic	67.3
American Indian or Alaskan Native, Non-Hispanic	42.6
Hispanic	39.7
Age	
3 to 15 years	49.7
16 to 44 years	68.9
45 to 64 years	63.2
65 years and over	33.4
Gender	
Male	59.3
Female	59.0
Household Type	
Married-couple with children	74.1
Single parent (male)	54.0
Single parent (female)	52.3
Family without children	59.5
Non-family household	50.2

Table Al. (Continued)

Disability Status	
Has a disability	33.1
No disability	65.1
Foreign-Born Status	
Citizens (including foreign born)	60.5
Non-Citizen	42.3
Sample Size	129,249
Estimated Population	289,420,157

Source: U.S. Census Bureau, Current Population Survey (CPS) and CPS School Enrollment and Internet Use Supplement, October 2009, and ESA calculations.

Note: People who report using the Internet at any location and live in a household that subscribes to a broadband Internet service are considered to use broadband Internet at home. Because of the way the CPS Internet Use Supplement questions are constructed, it is not possible to directly identify people who use broadband Internet at home. *Sample includes all people 3 years of age or older.

Table A2. Individual Broadband Internet Use by Gaograohje Characteristics, 2009

Individual Broadband Internet Use: Percent of persons connecting to the Internet at home using broadband	
All Persons*	59.1
Metropolitan Status	
Urban	61.1
Rural	48.5
Metropolitan Area (CBSA) Size	
Under 1,000,000	59.3
1,000,000-2,499,999	61.6
2,500,000-4,999,999	65.6
5,000,000 or more	60.1
Sample Size	129,249
Estimated Population	289,420,157

Source: U.S. Census Bureau, Current Population Survey (CPS) and CPS School Enrollment and Internet Use Supplement, October 2009, and ESA calculations.

Note: *Sample includes all people 3 years of age or older.

The procedure used in developing estimates for the entire civilian non institutional population for the CPS involves weighting sample results using independent estimates of the population by state, sex, age, race, and Hispanic/non-Hispanic categories. These independent estimates are developed by the Census Bureau using civilian non institutional population counts from the last decennial census and projecting them forward to current years using data on births, deaths, and net migration.

Section A2. Broadband Internet Use by Demographic Characteristics, 2009

This section provides data on person-level broadband Internet use by demographic and geographic characteristics. The report presents and analyzes broadband Internet adoption data at the household level. The underlying trends do not change if persons, as opposed to households and heads of households, are the unit of analysis.

Section A3. Marginal Effects of Demographic and Geographic Characteristics on the Likelihood that a Household Uses Broadband Internet at Home, 2009

This section contains the underlying regression results for Table 6 in Section 4.2. Section 4.2 indicates that we utilized a regression analysis framework to estimate the simultaneous impact of multiple factors on the probability that a household adopts broadband Internet services at home. The results allow us to isolate the effect of any one factor, holding all the other factors constant. We refer to these results as the marginal effects of selected demographic and geographic characteristics on home broadband Internet use.

Table A3. Marginal Effects: Regression of Broadband Internet Adoption on Demographic and Geographic Characteristics. 2009

Linear Probability Model	Column 1	Column 2
Family Income: Less than $25,000	omitted	omitted
Family Income: $25,000-$50,000	0.1597*** (0.007)	0.1593*** (0.007)
Family Income: $50,000-$75,000	0.2730*** (0.008)	0.2720*** (0.008)
Family Income: $75,000-$100,000	0.3141*** (0.008)	0.3122*** (0.008)

Table A3. (Continued)

Linear Probability Model	Column 1	Column 2
Family Income: $100,000 or more	0.3377*** (0.008)	0.3343*** (0.008)
Education: Less than High School Degree	omitted	omitted
Education: high School Degree	0.1066*** (0.008)	0.1065*** (0.008)
Education: Some College	0.2282*** (0.009)	02279*** (0.009)
Education: College Degree or more	0.2887*** (0.009)	0.2872*** (0.009)
Age	0.0067*** (0.001)	0.0068*** (0.001)
Age-squared	-0.0001*** (0.000)	-0.0001*** (0.000)
Non-Hispanic White	omitted	omitted
Non-Hispanic Black	-0.1002*** (0.008)	-0.1039*** (0.008)
Hispanic	-0.1407*** (0.009)	M.1429*** (0.009)
Non-Hispanic Asian	-0.0043 (0.011)	-0.0079 (0.011)
Non-Hispanic, Other	-0.0486*** (0.017)	ˉ0.0492*** (0.017)
Metropolitan (Urban)	0.0689*** (0.007)	
Nonmetropolitan (Rural)	omitted	omitted
Not-identified (as metropolitan or nonmetropolitan)	0.0463* (0.026)	
Metropolitan size: Less than 1 million		0.0636*** (0.007)
Metropolitan size: 1 to 2.5 million		0.0727*** (0.008)
Metropolitan size: 2.5 to 5 million		0.0908*** (0.009)
Metropolitan size: 5 million or more		0.0855*** (0.010)
Metropolitan size - not identified		0.0302** (0.012)

Linear Probability Model	Column 1	Column 2
Disability	-0.0525*** (0.007)	-0.0520*** (0.007)
Disability-not identified	0.0870*** (0.023)	0.0924*** (0.024)
Foreign-born non-citizen	-0.0564*** (0.011)	-0.0586*** (0.011)
Total number of persons in household	0.0255*** (0.002)	0.0258*** (0.002)
State indicator variables	yes	yes
Constant	0.0867*** (0.029)	0.0888*** (0.029)
Sample Size	43,662	43,662
Estimated Number of Households	94,963,684	94,963,684
R-squared	0.308	0.309

Robust standard errors in parentheses

*** $p<0.01$, ** $p<0.05$, * $p<0.1$

Source: U.S. Census Bureau, Current Population Survey (CPS) and CPS School Enrollment and Internet Use Supplement, October 2009, and ESA calculation

Note: The marginal effects reported on Table 6 in Section 4.2 are from the first column.

Section A4. Marginal Effects of Demographic and Geographic Characteristics on the Likelihood that a Household Uses Broadband Internet at Home, by Household Type, 2009

We performed our regression analysis separately for different household types in order to analyze how stable or robust the associations are between broadband Internet use and household attributes. Table A4 shows the marginal effects of demographic and geographic characteristics on the likelihood of broadband Internet use for four different household types: married-couple households with children, single-parent households with children, family households without children, and non-family households.

In this section, we will briefly discuss the results for married-couple households with children and single-parent households with children. Married-couple families with children represent 22% of our sample and single-parent families with children represent another 11%. The first column of Table A4 presents the results for the overall sample and the next two columns present the results for the two sub-groups.

The marginal effects of most of the characteristics are similar for our two family types, implying that the association between household attributes and broadband Internet use are rather robust among these two family types. For example, income has very similar effects on broadband Internet use for these two groups. The likelihood of broadband Internet use rises with income, and the rising effect of income diminishes as income grows. The effect of education is similar, except that a college degree (or more) has a steeper impact on single-parent families than their married-couple counterparts. Relative to households where the householder has less than a high school degree, a single-parent household headed by someone with a college degree is 31 percentage points more likely to use broadband Internet at home (all else similar), compared to 24 percentage points for their married- couple counterparts.

The adoption gap between White and Black households is larger for single-parent (11 percentage points) families than among their married-couple counterparts (6 percentage points). The adoption gaps between White and Hispanic households are similar for both married-couple and single-parent households.

The marginal effects of foreign-born status, disability status, metropolitan status, and household size are similar for the two groups.

Table A4. Marginal Effects: Regression of Broadband Internet Adoption on and Geographic Characteristics, by Household Type, 2009

Linear Probability Model	All	Married-couple with Children	Single-parents with Children	Family Households without Children	Non-family Household s
Family Income: Less than $25,000	omitted	omitted	omitted	omitted	omitted
Family Income: $25,000-$50,000	0.1597*** (0.007)	0.1552*** (0.019)	0.1404*** (0.019)	0.1405*** (0.014)	0.1207*** (0.011)
Family Income: $50,000-$75,000	0.2730*** (0.008)	0.2524*** (0.019)	0.2508*** (0.024)	0.2414*** (0.015)	0.2160*** (0.013)
Family Income: $75,000-$100,000	0.3141*** (0.008)	0.2718*** (0.020)	0.3195*** (0.031)	0.2912*** (0.016)	0.2695*** (0.016)
Family Income: $100,000 or more	0.3377*** (0.008)	0.2963*** (0.019)	0.3188*** (0.026)	0.3258*** (0.015)	0.2877*** (0.015)

Linear Probability Model	All	Married-couple with Children	Single-parents with Children	Family Households without Children	Non-family Households
Education: Less than High School Degree	omitted	omitted	omitted	omitted	omitted
Education: High School Degree	0.1066*** (0.008)	0.1429*** (0.021)	0.1102*** (0.024)	0.1094*** (0.016)	0.0830*** (0.012)
Education: Some College	0.2282*** (0.009)	0.2282*** (0.021)	0.2331*** (0.024)	0.2267*** (0.016)	0.2348*** (0.014)
Education: College Degree or more	0.2887*** (0.009)	0.2425*** (0.021)	0.3113*** (0.028)	0.2683*** (0.016)	0.3524*** (0.014)
Age	0.0067*** (0.001)	0.0126*** (0.003)	0.0055 (0.003)	0.0075*** (0.002)	-0.0026** (0.001)
Age-squared	0.0001*** (0.000)	-0.0001*** (0.000)	-0.0001 (0.000)	-0.0001*** (0.000)	-0.0000*** (0.000)
Non-Hispanic White	omitted	omitted	omitted	omitted	omitted
Non-Hispanic Black	0.1002*** (0.008)	-0.0568*** (0.018)	-0.1118*** (0.021)	-0.0868*** (0.016)	-0.1051*** (0.013)
Hispanic	0.1407*** (0.009)	-0.1051*** (0.015)	-0.1258*** (0.025)	-0.1423*** (0.016)	-0.1790*** (0.017)
Non-Hispanic Asian	0.0043 (0.011)	0.0226 (0.016)	0.0255 (0.055)	-0.0326 (0.020)	-0.0291 (0.023)
Non-Hispanic Other	0.0486*** (0.017)	-0.0943*** (0.034)	-0.0508 (0.041)	-0.0015 (0.031)	-0.0486* (0.028)
Disability	0.0525*** (0.007)	-0.0453* (0.024)	-0.0952*** (0.027)	-0.0185 (0.012)	-0.0577*** (0.011)
Disability - not identified	0.0870*** (0.023)	0.0879*** (0.031)	-0.1961 (0.165)	0.1368*** (0.024)	0.1282** (0.054)
Foreign-born non-citizen	0.0564*** (0.011)	-0.0818*** (0.019)	-0.1051*** (0.029)	-0.0521** (0.021)	-0.0119 (0.020)
Total number of persons in household	0.0255*** (0.002)	0.0001 (0.004)	0.0084 (0.006)	0.0296*** (0.006)	0.0430*** (0.008)
Metropolitan (Urban)	0.0689*** (0.007)	0.0704*** (0.013)	0.0901*** (0.023)	0.0703*** (0.011)	0.0630*** (0.012)
Nonmetropolitan (Rural)	omitted	omitted	omitted	omitted	omitted
Not-identified (as metropolitan or nonmetropolitan)	0.0463* (0.026)	0.1082** (0.042)	-0.0189 (0.099)	0.0541 (0.043)	-0.0091 (0.051)

Table A4. (Continued)

Linear Probability Model	All	Married-couple with Children	Single-parents with Children	Family Households without Children	Non-family Household s
Constant	0.0867*** (0.029)	0.0208 (0.079)	-0.0205 (0.090)	0.1162** (0.053)	0.3381*** (0.051)
Sample Size	43,662	9,857	4,551	14,673	14,581
Estimated Number of Households	94,963,684	21,734,535	10,381,473	30,977,040	31,870,635
R-squared	0.308	0.258	0.234	0.256	0.343

Robust standard errors in parentheses

*** $p<0.01$, ** $p<0.05$, * $p<0.1$

Source: U.S. Census Bureau, Current Population Survey (CPS) and CPS School Enrollment and Internet Use Supplement, October 2009, and ESA calculations.

Section A5. People with Disabilities: Profile and Internet Use, 2009

This section provides data on demographic and geographic characteristics for individuals with disabilities, and on Internet use by this group. The report presents and analyzes the data at the household level. The underlying trends do not change if persons, as opposed to households and heads of households, are the unit of analysis.

Table A5. Person Demogranhic and Geographic by Disability Status, 2009

	Distribution for People with disabilities (%)	Distribution for People with no disability (%)
Percent of All People	11.2	88.8
Family Income		
Less than $25,000	37.7	15.7
$25,000-$50,000	21.9	21.1
$50,000-$75,000	11.2	16.2
$75,000-$100,000	4.4	10.2
More than $100,000	5.1	16.2
Education		
Less than High School Degree	27.3	16.5

	Distribution for People with disabilities (%)	Distribution for People with no disability (%)
High School Degree	35.2	28.6
Some College	23.4	27.3
College Degree or more	14.2	27.6
Age (mean years)	60.1	42.6
Geographic Location		
Urban	78.2	84.5
Rural	21.1	14.8
Sample Size	12,638	94,573
Estimated Population	27,024,985	213,523,194

Source: U.S. Census Bureau, Curren Population Survey (CPS) and CPS School Enrollment and Internet Use Supplement, October 2009, and ESA calculations.

Note: The distributions across the income categories do not sum to 100% since income data are not reported by some households.

Table A6. Internet Use by Disability Status, 2009

	People with disabilities	People with no disability
Internet Use (%)		
At any location	41.4	74.8
At home	37.4	69.3
Use broadband Internet at home	33.1	65.1
Use dial-up at home	4.0	3.9
Sample Size	12,638	94,573
Estimated Population	27,024,985	213,523,194

Source: U.S. Census Bureau, Current Population Survey (CPS) and CPS School Enrollment and Internet Use Supplement, October 2009, and ESA calculations.

Section A6. Broadband Internet Use by State, 2001 versus 2009

This section presents average broadband Internet usage rates by state, as well as the standard errors and 90% confidence intervals for the estimated state-level average broadband Internet adoption rates.

Table A7. Average Broadband Internet Use by State, 2009

Adoption Rate (%)				
State	Mean	Standard Error	90% Confidence Interval	
			Lower Bound	Upper Bound
AL	48	2	45	52
AK	73	2	70	76
AZ	67	2	64	70
AR	51	2	48	54
CA	68	1	66	69
CO	69	1	67	71
CT	71	1	69	73
DE	67	2	64	69
DC	66	2	63	69
FL	67	1	65	68
GA	64	1	62	66
HI	70	2	67	73
ID	67	2	64	71
IL	63	1	61	65
IN	56	2	54	59
IA	62	2	60	65
KS	67	2	64	70
KY	54	2	51	57
LA	57	2	53	61
ME	61	1	59	64
MD	70	1	68	72
MA	73	2	70	75
MI	62	1	60	65
MN	67	1	65	69
MS	42	2	38	45
MO	57	2	55	60
MT	58	2	55	62
NE	64	2	61	67
NV	68	2	65	70
NH	73	1	71	75
NJ	72	1	70	75
NM	55	2	51	58
NY	66	1	64	67

Adoption Rate (%)				
State	Mean	Standard Error	90% Confidence Interval	
			Lower Bound	Upper Bound
NC	59	1	57	62
ND	63	2	60	66
OH	61	1	59	64
OK	56	2	53	59
OR	70	2	67	73
PA	62	1	60	64
RI	69	1	67	72
SC	53	2	50	56
SD	60	2	57	62
TN	55	2	52	58
TX	60	1	58	61
UT	73	2	70	76
VT	61	2	58	63
VA	65	1	63	67
WA	72	1	70	75
WV	52	2	49	55
WI	67	1	65	69
WY	66	2	63	68

Source: U.S. Census Bureau, Current Population Survey (CPS) and CPS School Enrollment and Internet Use Supplement, October 2009, and ESA calculations.

Table A8. Average Broadband Internet Use by State, 2001

Adoption Rate (%)				
State	Mean	Standard Error	90% Confidence interval	
			Lower Bound	Upper Bound
AL	5	1	4	7
AK	13	1	11	15
AZ	12	1	10	14
AR	4	1	3	5
CA	13	1	12	14
CO	10	1	9	12
CT	11	1	10	13
DE	6	1	5	8

Table A8. (Continued)

Adoption Rate (%)				
State	Mean	Standard Error	90% Confidence interval	
			Lower Bound	Upper Bound
DC	7	1	6	9
FL	10	1	9	11
GA	7	1	6	9
HI	18	2	16	21
II)	6	1	5	8
IL	6	1	5	7
IN	4	1	3	5
IA	9	1	7	10
KS	10	1	8	12
KY	3	1	2	4
LA	5	1	4	7
ME	10	1	8	11
MD	9	1	7	9
MA	14	1	13	16
MI	9	1	8	10
MN	8	1	7	9
MS	6	1	4	7
MO	7	1	6	9
MT	3	1	2	5
NE	10	1	8	12
NV	7	1	6	8
NH	14	1	12	16
NJ	12	1	11	14
NM	2	1	1	3
NY	12	1	11	13
NC	7	1	6	8
ND	5	1	3	6
OH	8	1	6	9
OK	8	1	6	9
OR	9	1	8	11
PA	8	1	7	9
RI	11	1	9	12
SC	8	1	7	10

State	Mean	Standard Error	90% Confidence interval	
			Lower Bound	Upper Bound
TN	10	1	8	12
TX	9	1	8	10
UT	12	1	10	14
VT	8	1	7	9
VA	8	1	6	9
WA	11	1	9	12
WV	4	1	3	6
WI	6	1	5	7
WY	6	1	4	7

Source: U.S. Census Bureau, Current Population Survey (CPS) and CPS Computer and Internet Use Supplement, September 2001, and ESA calculations.

Table A9. Marginal Effects: Regression of Home Internet Use on Demographic and Geographic Characteristic, 2001 and 2009

Linear Probability Model	2001	2009
Family Income: Less than \$25,000	omitted	omitted
Family Income: \$25,000-\$50,000	0.1550*** (0.006)	0.1843*** (0.007)
Family Income: \$50,000-\$75,000	0.2959*** (0.008)	0.2954*** (0.007)
Family Income: \$75,000 or more	0.3730*** (0.008)	0.3200*** (0.007)
Education: Less than High School Degree	omitted	omitted
Education: High School Degree	0.0971*** (0.007)	0.1320*** (0.009)
Education: Some College	0.2142*** (0.008)	0.2503*** (0.009)
Education: College Degree or more	0.2909*** (0.008)	0.3049*** (0.009)
Age	0.0074*** (0.001)	0.0103*** (0.001)
Age-squared	-0.0001*** (0.000)	-0.0001*** (0.000)

Table A9. (Continued)

Linear Probability Model	2001	2009
Non-Hispanic White	omitted	omitted
Non-Hispanic Black	-0.1631*** (0.008)	-0.1027*** (0.008)
Hispanic	-0.1487*** (0.010)	-0.1396*** (0.009)
Non-Hispanic Asian	0.0456*** (0.014)	-0.0205* (0.011)
Non-Hispanic, Other	-0.1191*** (0.027)	-0.0538*** (0.016)
Foreign-born non-citizen	-0.0623*** (0.011)	-0.0383*** (0.010)
Total number of persons in household	0.0358*** (0.002)	0.0288*** (0.002)
Metropolitan (Urban)	0.0361*** (0.006)	0.0500*** (0.006)
Nonmetropolitan (Rural)	omitted	omitted
Not-identified (as metropolitan	0.0212	0.0224
or nonmetropolitan)	(0.034)	(0.025)
State indicator variables	yes	yes
Constant	-0.0414* (0.025)	0.0402 (0.028)
Sample Size	47,310	43,662
Estimated Number of Households	88,963,933	94,963,684
R-squared	0.324	0.315

Robust standard errors in parentheses
*** p<0.01, ** p<0.05, * p<0.1

Source: U.S. Census Bureau, Current Population Survey (CPS), CPS School Enrollment and Internet Use Supplement, October 2009, CPS and CPS Computer and Internet Use Supplement, September 2001, and ESA calculations.

Section A7. Marginal Effects of Demographic and Geographic Characteristics on the Likelihood that a Household Uses Internet at Home, 2001 versus 2009

This section contains the underlying regression results for Table 26 in Section 8.4.1. The analysis presented in Section 8.4.1 involved performing the regression analysis separately for 2001 and 2009 in order to analyze whether the marginal effects of selected demographic and geographic characteristics on household Internet use have changed over time.

Section A8. Marginal Effects of Demographic and Geographic Characteristics on the Likelihood that a Household Uses Broadband Internet at Home, 2007 versus 2009

This section contains the underlying regression results for Table 27 in Section 8.4.2 for comparing the marginal effects of selected demographic and geographic characteristics on home broadband Internet use between 2007 and 2009.

Table A10. Marginal Effects: Regression of Home Broadband Internet Use on Demographic and Geographic Characteristics, 2007 and 2009

Linear Probability Model	2007	2009
Family Income: Less than $25,000	omitted	omitted
Family Income: $25,000-$50,000	0.1297*** (0.007)	0.1671*** (0.007)
Family Income: $50,000-$75,000	0.2619*** (0.008)	0.2818*** (0.008)
Family Income: $75,000-$100,000	0.3309*** (0.009)	0.3233*** (0.008)
Family Income: $100,000 or more	0.3851*** (0.009)	0.3474*** (0.008)
Education: Less than High School Degree	omitted	omitted
Education: High School Degree	0.0812*** (0.008)	0.1104*** (0.008)
Education: Some College	0.2137*** (0.009)	0.2321*** (0.009)
Education: College Degree or more	02971*** (0.009)	0.2936*** (0.009)
Age	0.0026*** (0.001)	0.0068*** (0.001)
Age-squared	-0.0001*** (0.000)	-0.0001*** (0.000)

Table A10. (Continued)

Linear Probability Model	2007	2009
Non-Hispanic White	omitted	omitted
Non-Hispanic Black	-0.1058*** (0.008)	-0.1004*** (0.008)
Hispanic	-0.1198*** (0.009)	-0.1396*** (0.009)
Non-Hispanic Asian	0.0412*** (0.013)	-0.0036 (0.011)
Non-Hispanic, Other	-0.0522*** (0.017)	-0.0527*** (0.016)
Foreign-born non-citizen	-0.0516*** (0.011)	-0.0527*** (0.011)
Total number of persons in household	0.0180*** (0.002)	0.0259*** (0.002)
Metropolitan (Urban)	0.0811*** (0.007)	0.0698*** (0.007)
Nonmetropolitan (Rural)	omitted	omitted
Not-identified (as metropolitan	-0.0033	0.0472*
or nonmetropolitan)	(0.026)	(0.026)
Constant	0.0686** (0.030)	0.0734** (0.029)
Sample Size	42,481	43,662
Estimated Number of Households	91,153,697	94,963,684
R-squared	0.298	0.307

Robust standard errors in parentheses

*** $p<0.01$, ** $p<0.05$, * $p<0.1$

Source: U.S. Census Bureau, Current Population Survey (CPS) and CPS School Enrollment and Internet Use Supplement, October 2007 and October 2009, and ESA calculations.

REFERENCES

[1] Lyle, Elizabeth. (2010). *A Giant Leap and a Big Deal,* OBI Working Paper Series No. 2, Federal Communications Commission, Spring. http://download.broadband.gov/plan/fcc-omnibus-broadband-initiative-%28obi%29-working-report-giant-leap-big-deal-delivering-promise-of-equal-access-to-broadband-forpeople-with-disabilities.pdf

[2] (2010). Federal Communications Commission. *High-Speed Services for Internet Access: Status as of December 31, 2008,* February.

[3] Horrigan, John. (2009). *Home Broadband Adoption*, Pew Internet and American Life Project, Pew Research Center, June.

[4] Horrigan, John. (2010). *Broadband Adoption and Use in America,* OBI Working Paper Series No. 1, Federal Communications Commission, February.

[5] (2010). Office of Management and Budget (OMB). OMB Bulletin No. 04-03, *Update of Statistical Area Definitions and Additional Guidance on Their Uses.* Accessed June 23. http://www.whitehouse.gov/omb/bulletinsfy04_b04-03/

[6] (2010). *National Telecommunications and Information Administration* (NTIA). Digital Nation: 21st Century America's Progress Toward Universal Broadband Internet Access, February.

[7] Smith, Aaron. (2010). *Home Broadband 2010,* Pew Internet and American Life Project, Pew Research Center, August.

[8] (2001). U.S. Census Bureau. *Computer and Internet Use Supplement File Technical Documentation,* CPS-02. Current Population Survey, September. http://www.census.gov/apsd/techdoc/cps/cpsoct Ol.pdf

[9] (2009). U.S. Census Bureau. *School Enrollment and Internet Use Supplement File Technical Documentation, CPS-09.* Current Population Survey, October. http://www.census.gov/apsd/techdoc/cps/cpsoct09.pdf

[10] U.S. Census Bureau. *Metropolitan and Micropolitan Statistical Areas.* Accessed June 23, 2010 (a). http://www.census.gov/population

[11] (2008). U.S. Census Bureau. *American Community Survey.* "Table: Selected Social Characteristics in the United States." Accessed July 15, 2010 (b). *http://factfinder.census.gov/servlet/ADPTable?_bm=y8c-geo_id=01000US8cqr_name=ACS_2008_1YR_GOO_DP28c-ds_name=ACS_2008_1YR_G00_8c*-Jang=en8c_caller=geoselect8c-redoLog=true8c-format

[12] (1990). U.S. Department of Justice. *Current text of the Americans with Disabilities Act of 1990 incorporating the changes made by the ADA Amendments Act of 2008,* Americans with Disability Act of. Accessed July 15, 2010. http://www.ada.gov/pubs/adastatute08.htm

End Notes

[1] In the 2009 CPS Internet Use Supplement, a household with at least one of the following Internet services is considered to have broadband: DSL, cable modem, fiber optics, satellite, wireless (such as Wi-Fi), mobile phone or PDA, or some other broadband Internet connection (U.S. Census Bureau, 2009). It is not possible to identify in the survey which particular broadband service a household uses.

[2] Data on average broadband use at the individual level are presented in Appendix Section A2. The underlying trends do not significantly change if persons, instead of households, are the unit of analysis.

[3] A negligible fraction of households (0.4% in 2009) reported using the Internet from home using "something else", that is, something other than a dial-up telephone connection or broadband. As a result, the total share of households connecting to the Internet from home is 68.6% = 63.5% (using broadband) + 4.7% (using dial-up) + 0.4% (using "something else").

[4] In the CPS, a person with at least one of the following conditions is considered to have a disability: hearing impairment; blindness; impaired vision despite wearing glasses; physical, mental, or emotional condition that impairs the ability to concentrate, remember, or make decisions; difficulty in walking or climbing stairs; difficulty in dressing or bathing; physical, mental, or emotional condition that impairs the ability to do errands alone such as visiting a doctor's office or shopping (U.S. Census Bureau, 2009).

[5] American citizens include both native-born American citizens as well as foreign-born persons who are naturalized U.S. citizens.

[6] The geographic variable for identifying a household's location as urban or ruml is not available in the CPS public use files. This report uses the terms "urban" and "rural" to refer to metropolitan and nonmetropolitan areas, respectively. The definition of a metropolitan area (effective since 2000) is based on "core based statistical area" (CBSA), which includes both metropolitan and micropolitan statistical areas. According to the 2000 standards, each CBSA must have at least one urban area with at least 10,000 inhabitants. Each metropolitan statistical area must contain at least one urbanized area with population 50,000 or more. Each micropolitan statistical area must contain at least one urban duster with population between 10,000 and 50,000. As of June 6, 2003, there are 362 metropolitan statistical areas and 560 micropolitan statistical areas in the U.S. For more information, see U.S. Census Bureau (2010a) and Office of Management and Budget (2010).

[7] Table 6 and Figure 3 show the adoption gap between urban and rural households without controlling for urban area size. Figure 4 and Appendix Table A3 (column 2) show the urban-rural gap by urban area size.

[8] In the CPS, a person with at least one of the following conditions is considered to have a disability: hearing impairment; blindness; impaired vision despite wearing glasses; physical, mental, or emotional condition that impairs the ability to concentrate, remember, or make decisions; difficulty in walking or climbing stairs; difficulty in dressing or bathing; physical, mental, or emotional condition that impairs the ability to do errands alone such as visiting a doctor's office or shopping (U.S. Census Bureau, 2009).

[9] Changing the unit of analysis from household level to individual level does not change the underlying patterns (the person-level data are presented in Section A5 of the Appendix).

[10] According to the October 2009 Current Population Survey, which collected information on disability only for adult civilian household members, there were about 27 million adults with disabilities.

[11] Broadband refers to a Digital Subscriber Line (DSL) or a cable modem, the two dominant technologies of choice, in the 2001 CPS Internet Supplement data. In the 2009 CPS Supplement data, broadband refers to DSL, cable modem, fiber optics, satellite, wireless (such as Wi-Fi), mobile phone or PDA, or some other broadband Internet connection.

[12] Unlike the regression analysis for 2009 presented in Section 4.2, we are unable to control for disability status in this section since similar data on disability are not available for 2001.

In: Digital Nation: Broadband Internet Adoption… ISBN: 978-1-61324-556-9
Editors: Thomas N. Eaton et al.

Chapter 2

DIGITAL NATION: EXPANDING INTERNET USAGE

U.S. Department of Commerce

FOREWORD

"This isn't just about faster Internet or fewer dropped calls. It's about connecting every part of America to the digital age. It's about a rural community in Iowa or Alabama where farmers and small business owners will be able to sell their products all over the world. It's about a firefighter who can download the design of a burning building onto a handheld device; a student who can take classes with a digital textbook; or a patient who can have face-to-face video chats with her doctor."

–President Obama, State of the Union Address, January 25, 2011

The Internet is an extraordinary platform for innovation, economic growth, and social communication. It has become the backbone of a global digital infrastructure and an integral part of the information economy. Statistical reports on job growth and retail sales show the importance of the Internet economy. Between 1998 and 2008, the number of domestic information technology jobs grew 26 percent four times faster than employment in the United States as a whole. Moreover, global online transactions currently total an estimated $10 trillion annually.

As the Internet economy grows, the nature and pace of its growth depend on the ability of the nation's citizenry to adopt and make use of the Internet. Supply side factors, such as availability of broadband service, are critical. Equally important, however, is the ability of consumers to participate in online opportunities. The rapid diffusion of broadband Internet in the United States relative to other major technologies over time underscores how important this infrastructure is to Americans.

How is the United States faring? The U.S. Census Bureau's Current Population Survey (CPS) data documents the continuing rise in broadband Internet adoption among Americans, particularly at home. Based on the survey of 54,000 households and 129,000 persons, today more than 68 percent of U.S. households use high-speed broadband access. At the same time, dial-up use continues to wane as Americans migrate in large numbers to high-speed access and the substantial benefits it can confer. The wave of broadband Internet adoption cuts across all demographic groups, in rural and urban America alike.

There has been good progress, yet almost one-third of American households still lack a broadband connection. We need to make sure no one is left behind in the digital age. Significant gaps in Internet usage still exist among certain demographic and geographic groups around the country. People with college degrees adopt broadband at almost triple the rate of those with some high school education (84% versus 30%), among adults 25 years and older. The rates for White (68%) and Asian non-Hispanics (69%) exceed those for Black non-Hispanics (50%) and Hispanics (45%) by 18 percentage points or more. Rural America lags behind urban areas by ten percentage points (60% versus 70%).

Recognizing these disparities, the President recently announced a National Wireless Initiative. The goal of this initiative is to ensure that virtually all Americans (at least 98%) have access to high-speed wireless services (4G) within the next five years. The President's initiative represents an extraordinary opportunity to bring cost-effective, mobile broadband connectivity to areas currently without any broadband service. The initiative will bridge the existing gaps in broadband deployment, and will simultaneously serve as a catalyst for our economy creating new jobs, new investment, and incentives to develop new technologies.

Moreover, the President's wireless initiative also addresses another key issue raised by the survey, namely, that a significant number of households do not subscribe to broadband because they have inadequate computers or no computers at all. The initiative will help eliminate this access barrier by expanding 4G and facilitating the proliferation of mobile devices with computing capability. These mobile devices are significantly less expensive than computers, but still provide a means through which individuals can access the Internet, web applications, and other web services.

In addition, policymakers continue to focus on other ways to expand broadband usage in the United States. In particular, the American Recovery and Reinvestment Act of 2009 established broadband grant and loan programs through the Broadband Technology Opportunities Program (BTOP) and the Broadband Infrastructure Program (BIP), funding infrastructure projects, public computing centers, and sustainable adoption programs in many communities across the country. Each of these programs aims to close the deployment and adoption gaps documented in this report. Similarly, the Federal Communications Commission (FCC) recently released a Notice of Proposed Rulemaking on ways to expand the current universal service program to include broadband service.

At the same time, the Administration, working in close coordination with the States, is collecting data on broadband at an unprecedented level. For example, the National Telecommunications and Information Administration (NTIA), in collaboration with the FCC, developed the interactive National Broadband Map with data provided by the States and other stakeholders as part of the State Broadband Data and Development (SBDD) grant program. The map shows broadband deployment across the United States and provides specific information about the available broadband technologies and speeds.

The data collected from these programs, including data collected for the CPS survey, is essential to enabling the President's wireless initiative and assisting the FCC with universal service reform. These efforts will also help state and local policymakers throughout the country, in cooperation with private partners, to make sound, fact-based decisions in this area that will profoundly impact America's world competitiveness and standard of living.

Finally, consistent with this Administration's directive to maintain a transparent and open government, we will make the raw survey data utilized in this report publicly available for use and analysis by all interested parties.

Lawrence E. Strickling
Assistant Secretary for Communications and Information
U.S. Department of Commerce

Acknowledgments

The project team would like to thank Tom Power, Kathy Smith, Josephine Arnold, Dennis Amari, and Catherine Schmierer of NTIA; and Beethika Khan of the Economics and Statistics Administration for their contributions to this report.

Executive Summary

The new millennium has witnessed a rapid rise in broadband Internet connections in the United States. While the growth rate of total Internet connections at home slowed somewhat after 2001, the number of household broadband connections grew rapidly over the last decade. As of October 2010, the nationwide broadband adoption rate equaled 68.2 percent of households, up from 63.5 percent one year earlier. Overall, the growth of broadband Internet adoption in the United States stems from two sources: (1) supplanting dial-up services in many households, and (2) expanding the market for household Internet connections.

The nationwide trend toward widespread adoption of broadband Internet connections has occurred among virtually all demographic groups, with nearly universal year-over-year growth. Adoption and growth rates, however, are uneven among various demographic groups. Over the past year, gaps decreased in a number of demographic categories, but persisted with respect to family income, education, age, race/ethnicity, employment status, household type, and disability status. Gender-specific adoption rates remained at virtual parity. The average urban-rural differential narrowed, but a ten-percentage point gap remains. State-specific adoption varied widely. Almost 72 percent of Americans use the Internet at some location, increasing from 68.4 percent twelve months prior. The most popular locations for Internet access outside the home are the workplace and school, followed by public libraries and "someone else's house."

Nationwide, non-adoption at home for the Internet and broadband stands at 28.9 percent and 31.8 percent of households, respectively. The proportion of all persons who do not use the Internet anywhere equals 28.3 percent. For households that do not connect to broadband, the reason given most frequently for non-adoption was "don't need/not interested," followed by "too expensive." This ranking holds for the main sub-category of this group (i.e., those who do not use the Internet anywhere), but the cost factor ranks first for non-broadband households that either only use the Internet outside the home, or only connect through dial-up. Cost also becomes more important in such decisions for lower-income, and Black non-Hispanic and Hispanic households. Urban and rural households both rank "don't need," "too expensive," and "no/inadequate computer" as their top three reasons for non-adoption, but lack of broadband availability was much more significant in rural areas (9.4%) than urban locales (1.0%).

INTRODUCTION

By a variety of measures, broadband Internet is an essential part of this nation's economic fabric.[1] The rate of broadband adoption has outstripped such staples as VCRs, cell phones, cable, color televisions, personal computers, radios, electricity, and telephones.[2]

Is there a basis for society to promote the deployment and adoption of broadband Internet? Although research is in its early stages, the available evidence supports the notion that broadband can create significant economic benefits. For example, in 2009, Professor Leonard Waverman concluded:

> Research suggests that with the right skills and infrastructure in place, broadband strategies could increase national productivity and growth by up to 15%. This productivity improvement will increase GNP without increasing resources used in production. For example, the U.S. could increase its GDP by $100 billion with an increase of 10 additional broadband lines per 100 individuals.[3]

The same year, two Harvard Business School researchers estimated that Americans whose jobs related to the Internet contributed approximately $300 billion of economic activity to the United States' Gross Domestic Product. The researchers calculated that the direct and indirect economic value of the services that the Internet provided to the rest of the U.S. economy equaled $444 billion.[4] In addition, the Economic Development Administration (EDA) performed an econometric study of communities from 1998-2002, and found that the "results support[ed] the view that broadband access *does* enhance economic growth and performance, and that the assumed (and oft-touted) economic impacts of broadband are real and measurable."[5]

In the analysis below, NTIA examines the data collected in the CPS. The special Internet Use Supplement, attached to the CPS, periodically surveys approximately 54,000 households and gathers information on some 129,000 persons.[6] The Census Bureau conducted its ninth survey in October 2010.[7] This report documents the rapid growth of both broadband and the Internet in general, and examines how increases in adoption varied by demographic groups and geographic areas. Finally, it examines the major reasons why some Americans do not access broadband Internet at home. The raw data that underlies this report is available at http://www.ntia.doc.gov/data/index.html and can be found through dataset pointers at http://www.data.gov.[8]

ACCESS AND USE

Over the last decade, the number of household broadband Internet connections in the United States has risen rapidly. In August 2000, home broadband connections were quite rare, present in just 4.4 percent of households. Internet use in general had already become relatively popular, with connections in 41.5 percent of households, but the vast majority of Internet-using Americans relied on dial-up service at home. By October 2010, however, the landscape shifted dramatically, with 68.2 percent of households connecting via broadband, while 2.8 percent of households used dial-up service (see Figure 1).

As Figure 1 shows, the growth in broadband connections accelerated during most of the last decade, but the rate of growth in total Internet connections at home slowed after 2001. Overall, broadband Internet adoption in the United States traces its growth to two phenomena: (1) supplanting dial-up services in many households, and (2) expanding the market for household Internet connections. These numbers suggest that Americans have developed a strong preference for high-speed connections and the applications they enable. Although the gap between home broadband Internet adoption and telephone subscribership has narrowed over time, household use of broadband currently remains more than 25 percentage points behind.

The nationwide trend toward increased adoption of broadband Internet connections is observable among virtually all demographic groups, with year-over-year growth present almost universally. However, adoption and growth rates are uneven among various demographic groups, and such disparities warrant further investigation.

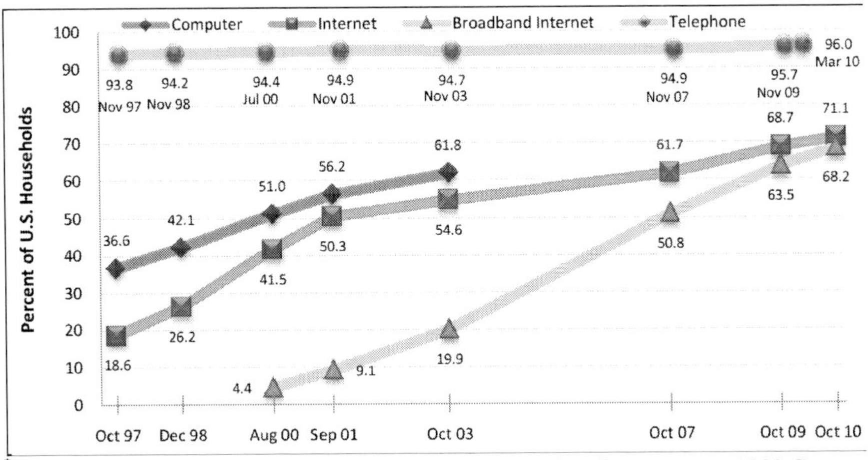

* Note: 2001-2010 use 2000 Census-*bazsed* weights and earlier years use 1990 Census-based weights.

Figure 1. Households with Computers, Telephone Subscriptions, and Internet Access, Selected Years, 1997-2010*.

Use by Family Income

Broadband use has consistently correlated positively with family income. Low income persons are less likely to use broadband at home than their more affluent peers, with the October 2010 data ranging from a 32.1 percent adoption rate among persons with $15,000 or less in annual family income, to 89.6 percent adoption among those with family incomes in excess of $150,000 (see Figure 2).

This trend in income-based adoption disparity appeared in the October 2009 and October 2007 data. A time-series comparison suggests that this form of digital divide is slowly shrinking. For example, broadband adoption in the $15,000-24,999 income range grew by 7.4 percentage points an impressive 21 percent increase between October 2009 and October 2010, while adoption in the $100,000-149,999 range increased by less than one percent during the same period. Similarly, higher growth among lower income groups also occurred between October 2007 and October 2009.[9]

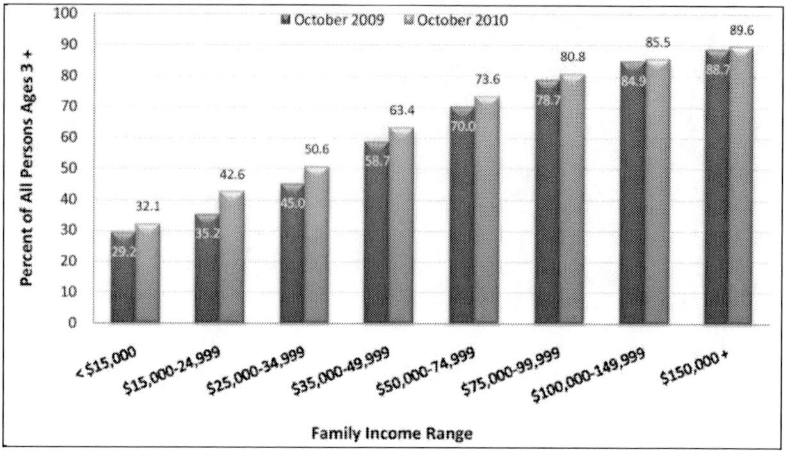

Figure 2. Persons Using Broadband in the Home by Family Income, 2009-2010.

Use by Education

Education is a particularly strong predictor of broadband use. While adults possessing at least a bachelor's degree are very likely to use broadband Internet at home (84.2%), individuals whose education ended with a high school diploma have a 51.4 percent adoption rate. Broadband Internet use drops even further among those without diplomas (see Figure 3).[10]

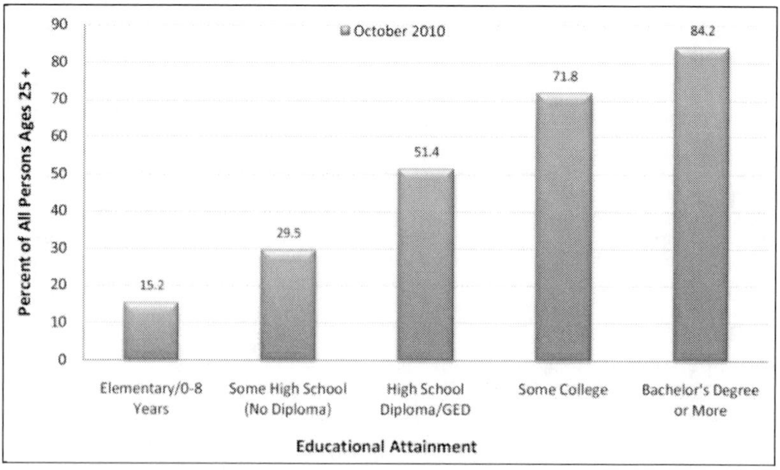

Figure 3. Persons Using Broadband in the Home by Education, 2010.

Use by Age

As was the case in the previous year, 18-24 year olds continued to show the highest rate of broadband use, while the lowest adoption rate was among persons ages 55 and older (see Figure 4). Seniors did exhibit one of the largest growth rates (4.0 percentage points), and the disparity in broadband use among older Americans compared with their younger peers narrowed slightly in the October 2010 data. The very modest year-over-year decline for 18-24 year olds, if statistically significant, was inconsistent with prevailing trends.

Use by Race and Ethnicity

The theme of persistent but slowly narrowing demographic disparities also holds for race and ethnicity (see Figure 5). Asian non-Hispanic Americans continued to lead in broadband use at home with a 68.8 percent adoption rate,[11] while White non-Hispanics have nearly closed a modest gap by growing to 68.3 percent adoption. Significant disparities, however, remained among other race and ethnic groups, with none exceeding broadband use of greater than 50 percent. Black non-Hispanics, American Indian/Alaskan Native non-Hispanics, and Hispanics are experiencing higher growth rates in adoption (similar to what NTIA observed in last year's comparison of 2007 and 2009 data), providing some evidence that these disparities are shrinking.

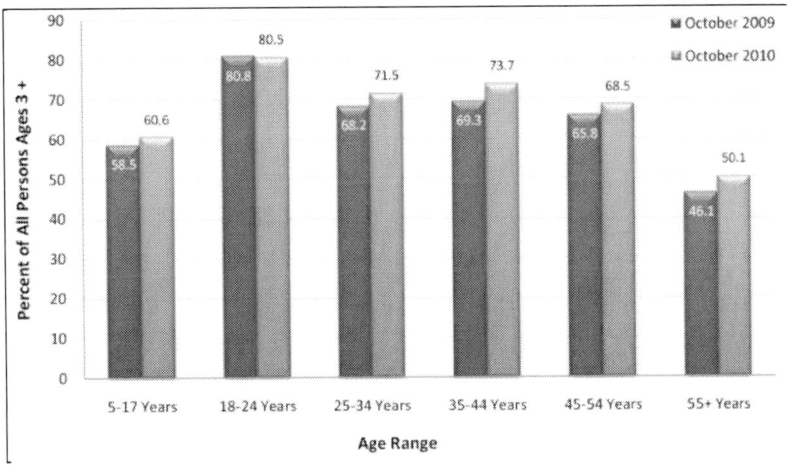

Figure 4. Persons Using Broadband in the Home by Age, 2009-2010.

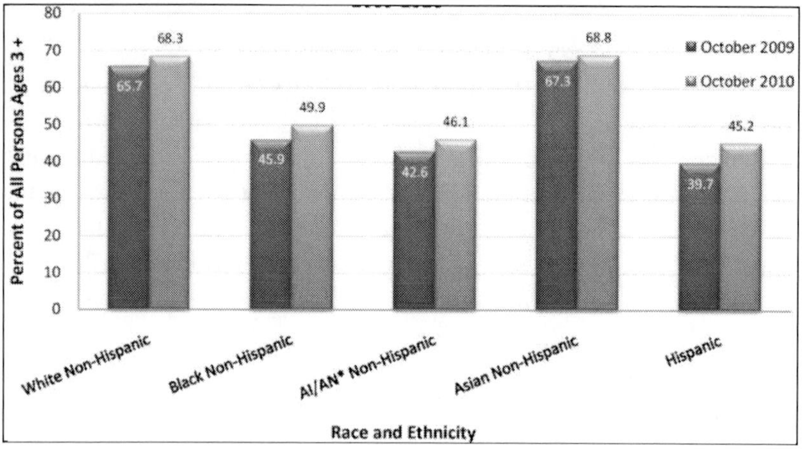

* American Indian/Alaskan Native

Figure 5. Persons Using Broadband in the Home by Race and Ethnicity, 2009-2010.

Use by Employment Status

While broadband use has increased among the employed, unemployed, and individuals who were not in the labor force, disparities between these groups persist (see Figure 6). As of October 2010, 73.4 percent of employed Americans were broadband users, compared with 61.9 percent of the unemployed and 50.5 percent of individuals not in the labor force. This category includes only individuals ages 16 and older, so that children who are not old enough to work do not skew the numbers.

Use by Household Type

Married couples with children less than 18 years of age retained the highest rate of household broadband adoption in 2010, while non-family households continued to lag behind all other groups (see Figure 7). Family households without children had the second highest incidence of broadband use, while single-parent households with children were in the middle of the range.

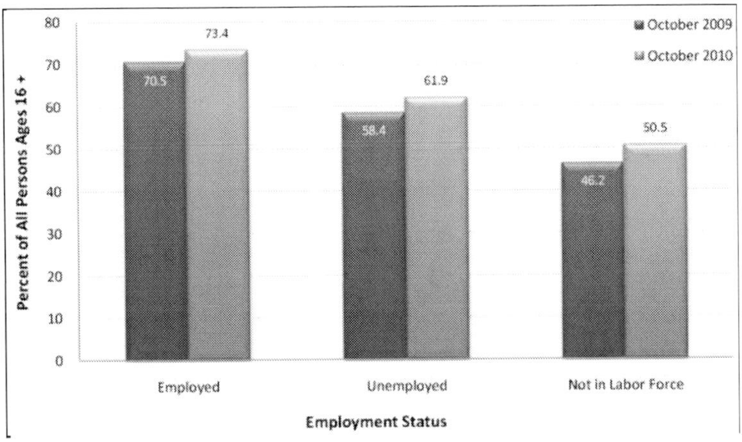

Figure 6. Persons Using Broadband in the Home by Employment, 2009-2010.

Use by Gender

The gap in broadband use between males and females continued to be very small, with home broadband use among 62.3 percent of males and 62.0 percent of females. This is consistent with the October 2009 data (see Figure 8). This parity between genders dates back to at least August 2000, when approximately 45 percent of both males and females responded that they used the Internet anywhere (i.e., inside or outside the home).

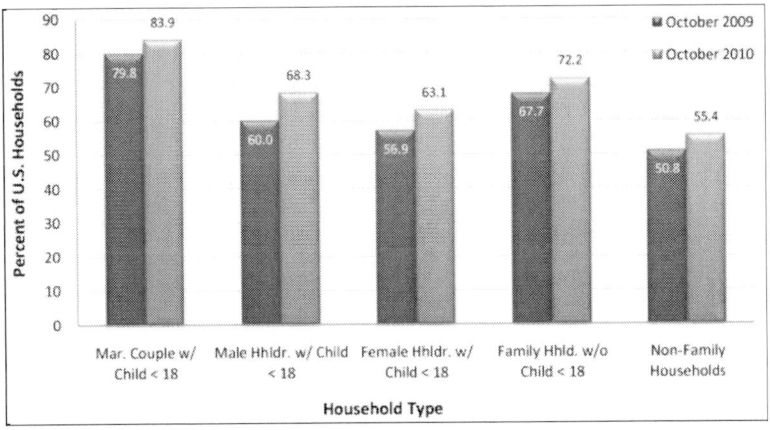

Figure 7. Households with Broadband Internet Connections at Home by Household Type, 2009-2010.

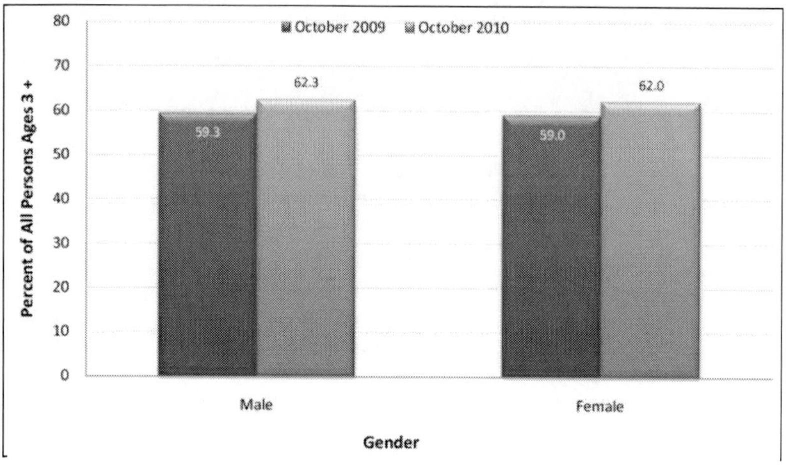

Figure 8. Persons Using Broadband in the Home by Gender, 2009-2010.

Use by Disability Status

In the November 2010 Commerce Department study, NTIA began tracking broadband adoption by disability or impairment status.[12] Based on October 2010 CPS data, Americans with disabilities are dramatically less likely to use broadband at home than the population without disabilities. While 68.0 percent of persons not reporting a disability use broadband at home, the adoption rate for persons with disabilities consistently lagged in a direct comparison (see Figure 9). Among the categories of persons with disabilities, the highest rate of broadband adoption occurred in those who are either deaf or have difficulty hearing (36.5%), followed by those with difficulty concentrating or remembering (35.3%). The adoption rates for each of the other disability categories were roughly less than half the adoption rate for those who reported no disability in October 2010.[13] As a group, 37.5 percent of persons with disabilities adopted broadband at home.

Use by Location

An urban-rural differential continued to exist in 2010 (see Figure 10). The differential, however, decreased during the past year, from 11.8 to 10.1

percentage points. Three years earlier, in 2007, the gap was 15.0 percentage points (53.8% versus 38.8%).

Beyond the differential based on population density, major differences in broadband use existed among the various states. An estimated 79.7 percent of households in Utah had broadband access, compared with 51.7 percent of households in Mississippi. The remaining states occupied the entire range in between (see Figure 11).

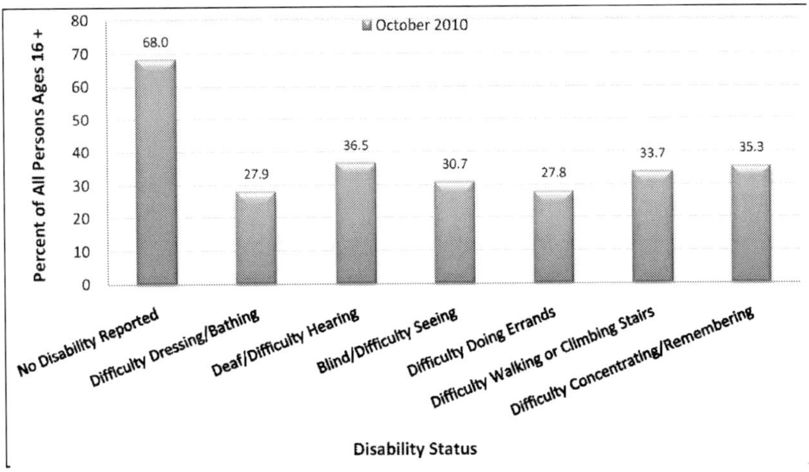

Figure 9. Persons Using Broadband in the Home by Disability Status, 2010.

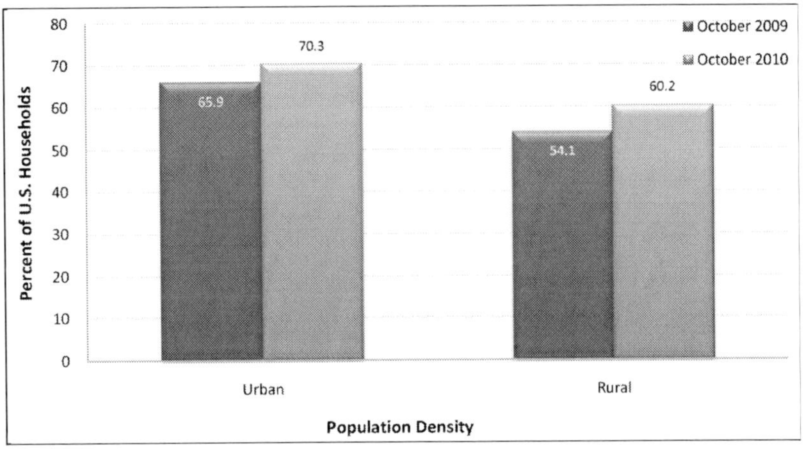

Figure 10. Households Using Broadband in the Home by Population Density, 2009-2010.

State	% w/ Broadband Internet	90% Confidence Interval		State	% w/ Broadband Internet	90% Confidence Interval	
		Upper Bound	Lower Bound			Upper Bound	Lower Bound
Utah	79.7	81.9	77.6	**Illinois**	68.7	70.3	67.2
New Hampshire	77.8	80.3	75.3	**Georgia**	68.6	70.4	66.8
Washington	76.7	78.6	74.8	**Delaware**	68.4	71.1	65.7
Massachusetts	75.9	77.9	73.9	**Iowa**	67.5	70.2	64.8
Connecticut	74.8	77.3	72.3	**Maine**	67.4	70.3	64.5
Oregon	74.7	77.2	72.2	**Pennsylvania**	67.4	69.0	65.8
Kansas	74.6	77.1	72.1	**Texas**	66.8	68.0	65.6
Nevada	74.2	76.7	71.7	**Michigan**	66.3	68.1	64.5
Arizona	74.2	76.3	72.1	**South Dakota**	65.5	68.0	63.0
Maryland	74.1	76.4	71.9	**North Carolina**	65.1	67.0	63.2
Alaska	73.4	76.0	70.8	**Missouri**	64.3	66.7	62.0
New Jersey	73.3	75.2	71.4	**Ohio**	63.9	65.6	62.2
California	73.1	74.0	72.2	**Oklahoma**	62.5	65.2	59.8
Wyoming	72.9	75.5	70.3	**Montana**	61.4	64.0	58.8
Idaho	72.0	74.5	69.5	**Louisiana**	60.5	63.2	57.8
District of Columbia	71.7	74.2	69.2	**South Carolina**	59.5	62.2	56.8
Colorado	71.6	74.0	69.2	**Tennessee**	59.5	61.8	57.2
North Dakota	70.9	73.4	68.4	**West Virginia**	59.1	61.7	56.5
Rhode Island	70.7	73.4	68.0	**Indiana**	58.8	61.1	56.5
Minnesota	70.6	72.9	68.3	**Kentucky**	57.8	60.6	55.0
Wisconsin	70.5	72.8	68.3	**New Mexico**	57.7	60.5	54.9
Florida	70.2	71.4	69.0	**Alabama**	55.5	58.2	52.9
Virginia	69.5	71.5	67.5	**Arkansas**	52.4	55.1	49.7
Vermont	69.2	72.0	66.4	**Mississippi**	51.7	54.5	48.9
Hawaii	69.2	71.9	66.5				
New York	69.0	70.3	67.7				
Nebraska	68.9	71.6	66.2				

Note: States are ordered by estimated household broadband usage rate for ease of understanding and not as a ranking. Rates should be understood in the context of their associated confidence intervals.

Figure 11. Households Using Broadband in the Home by State, 2010.

Internet Use anywhere

In 2010, the incidence of Internet use anywhere (i.e., inside or outside the home) by Americans totaled 71.7 percent (209.4 million persons, ages three and older). This represented an increase from 68.4 percent (197.9 million) in 2009. Similar to the broadband pattern, all demographic categories with respect to Internet use anywhere experienced rising adoption over time, but historical demographic differences in use have continued. The urban-rural gap in Internet use anywhere receded from 4.4 percentage points (69.3% versus 64.9%) in 2009, to 3.6 percentage points (72.4% versus 68.8%) in 2010.

Use Outside the Home

While our primary focus is on broadband use at home, Americans are also using the Internet in other locations.[14] The most popular locations for Internet access outside the home are the workplace (40.2%) and school (27.3%). Public libraries (11.3%) and "someone else's house" (9.0%) represent other important locations often utilized for this purpose (see Figure 12).

NON-ADOPTION

For policymakers, non-adoption of broadband Internet the obverse of this report's focus thus far merits special attention. The CPS data revealed that both non-adoption of Internet (28.9%) and broadband at home (31.8%) equaled less than one-third of U.S. households (see Figure 13). In 2007, the respective non-use rates were 37.6 percent and 49.2 percent. Both categories exhibited a decline during 2009-2010: no Internet use decreased by 2.4 percentage points; no broadband use decreased by 4.7 percentage points. Regarding non-adoption anywhere (i.e., neither at home nor outside the home), the incidence decreased from 31.6 percent of all persons three years old and older in 2009, to 28.3 percent in 2010.

Reasons for Non-Adoption

Thirty-eight million households in the United States do not access broadband Internet at home. In both 2009 and 2010, the reason given most frequently for non-adoption was "don't need/not interested," followed by "too expensive" (see Figure 14). With one exception, this ranking order held true first for the Internet beginning in December 1999, and later, broadband.[15] In 2010, lack of interest scored higher than ever before, at 45.6 percent, up from 37.8 percent the prior year. The cost factor decreased from 26.3 percent to 25.3 percent during the same interval. Home computer deficiencies ranked third, but also fell from 18.3 percent to 14.2 percent. Lack of availability also declined as a main reason, from 3.6 percent to 3.1 percent. By contrast, using broadband access somewhere else increased as a rationale, from 4.4 percent to 4.9 percent.

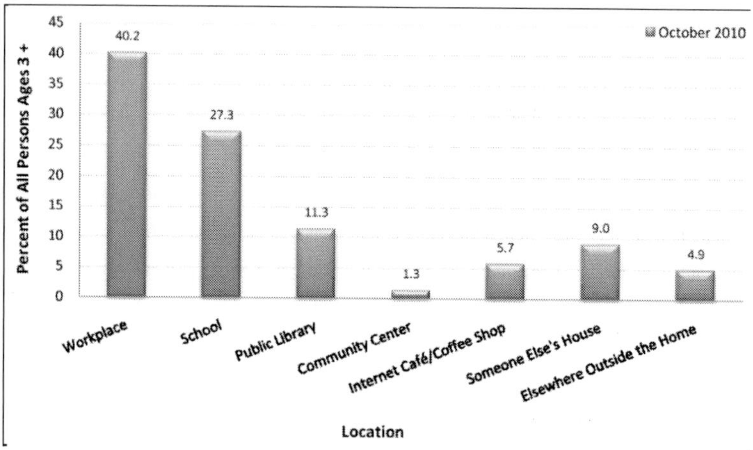

Figure 12. Persons Using Internet Outside the Home by Location, 2010.

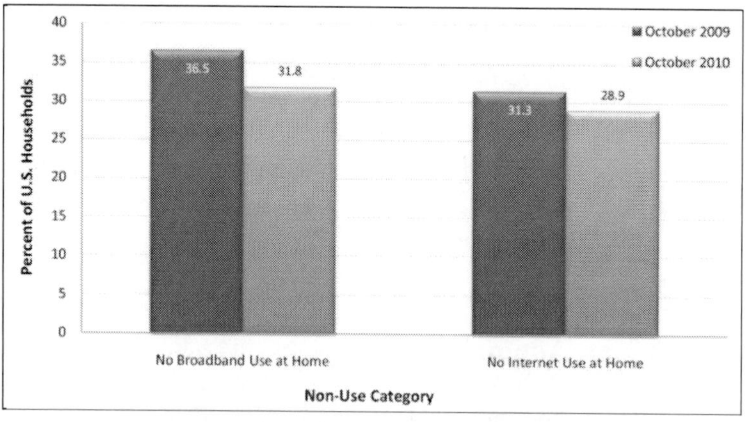

Figure 13. Households with No Internet Use and No Broadband Use at Home, 2009-2010.

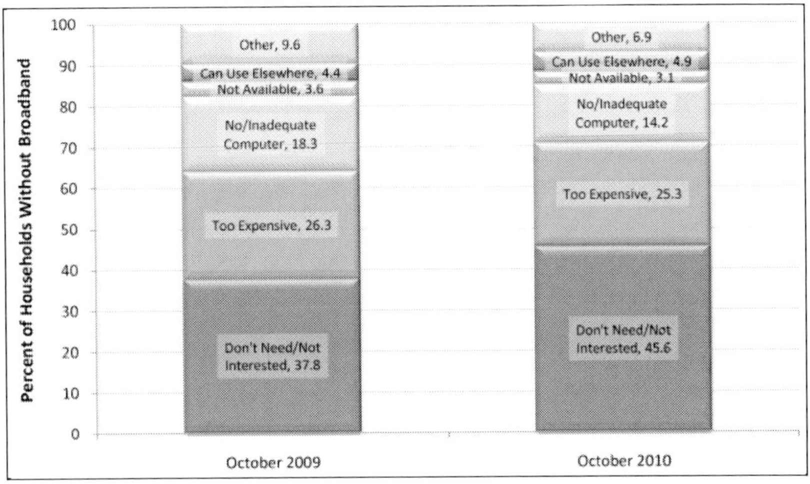

Figure 14. Main Reason for No High Speed Internet Use at Home, 2009-2010.

Demographic Differences in Non-Adoption

The next two tables (Figures 15 and 16) show the different reasons reported for not having broadband, based on family income and race. For family income, "too expensive" and "no/inadequate computer" became less prominent reasons as income increased, displaced in part by lack of availability and other choices. Budgetary and equipment concerns appeared to be important deterrents to broadband adoption among low-income households, while high-income households without broadband more likely lack interest or availability (see Figure 15).

Reasons for not using broadband also varied when examined by race. White and Asian non- Hispanics, who had the highest rates of broadband adoption, had non-users whose top-rated reason for eschewing broadband at home was, overwhelmingly, "don't need/not interested." Other racial or ethnic groups, however, were more likely to list "too expensive" or "no/inadequate computer," with those two responses exceeding 50 percent among Hispanics, compared with just over one-third of White non-Hispanics (see Figure 16).

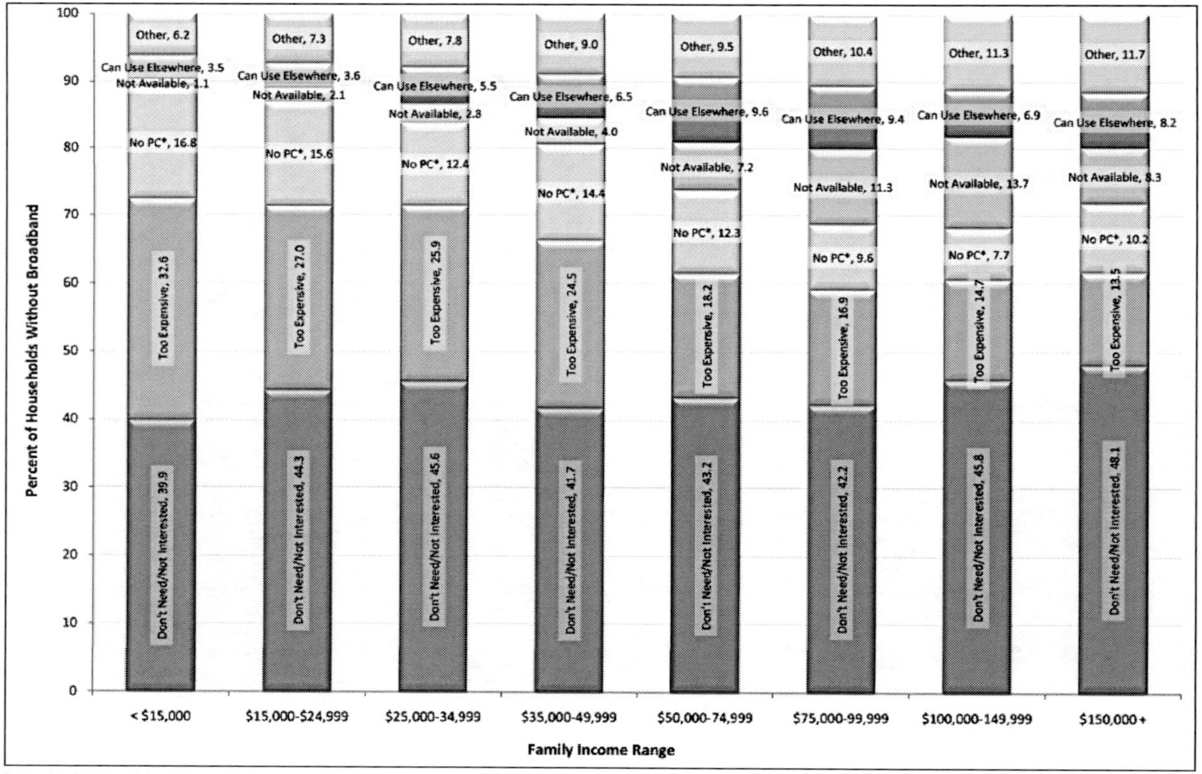

* Respondent indicated having either no computer at home or one inadequate for broadband Internet applications.

Figure 15. Main Reason for No High Speed Internet Use at Home by Family Income, 2010.

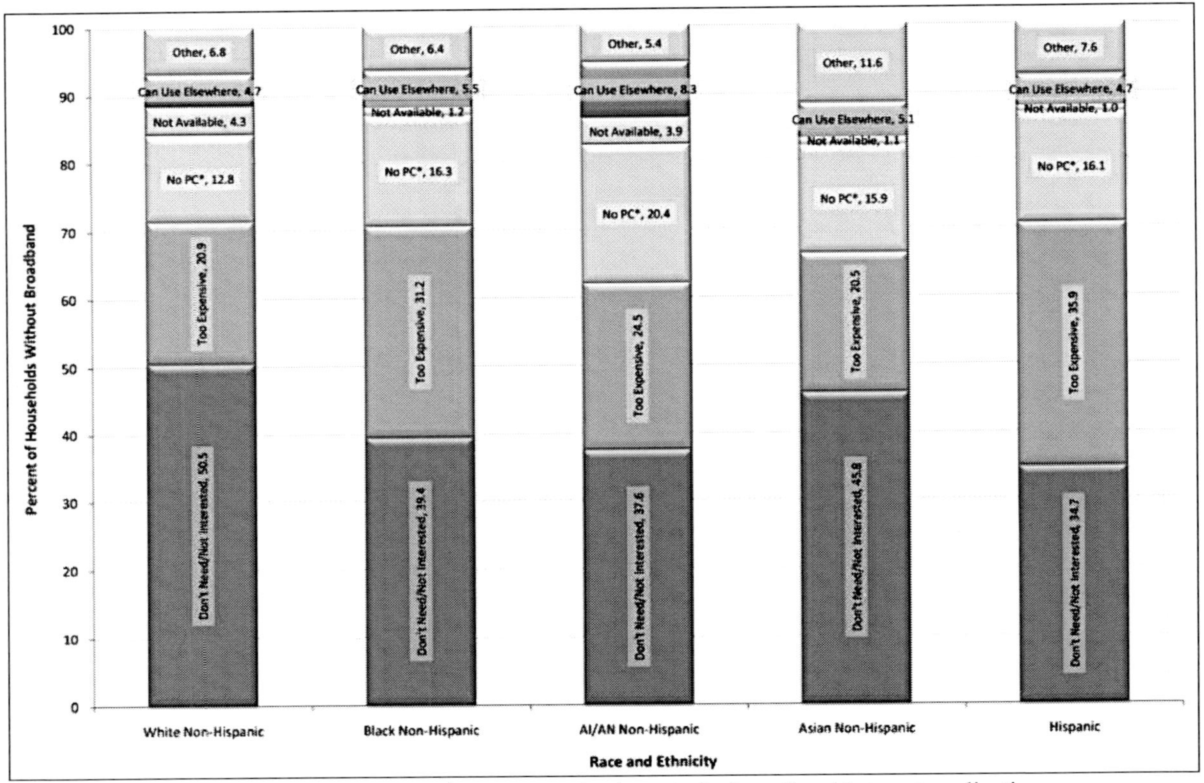

* Respondent indicated having either no computer at home or one inadequate for broadband Internet applications.

Figure 16. Main Reason for No High Speed Internet Use at Home by Race and Ethnicity, 2010.

Population density also affected reasons for non-adoption of broadband (see Figure 17). While the top-ranked reason for both urban and rural areas was "don't need/not interested" (45.4% versus 46.1%), respondents in urban areas cited cost more often than their rural counterparts (26.7% versus 21.1%). Computer deficiencies (14.6% versus 12.9%) and "can use elsewhere" (5.4% versus 3.7%) also weighed more heavily in urban locales. By contrast, lack of broadband availability, as a barrier in rural areas (9.4%), outranked the corresponding impact in urban America (1.0%).

Types of Broadband Non-Users

It is useful to disaggregate households lacking high-speed access at home into three sub-categories, thus permitting additional insights for policymakers: (1) households that do not use the Internet anywhere; (2) households that do not have an Internet connection, but had users outside the home; and (3) households with only a dial-up connection.

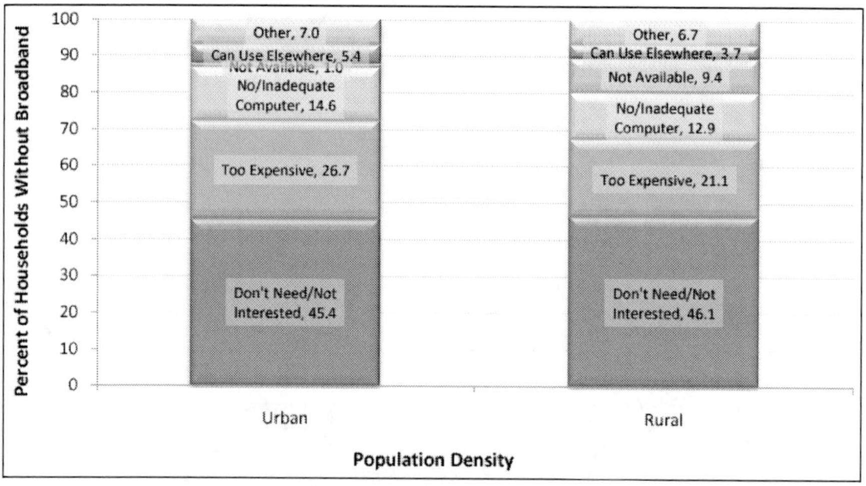

Figure 17. Main Reason for No High Speed Internet Use at Home by Population Density, 2010.

Among households lacking high-speed access at home, almost two-thirds (62.3 %) reported no Internet usage anywhere. "Don't need/not interested" (58.1%) far exceeded any other reason provided by these non-Internet users and rose almost 11 percentage points from the prior year, when it stood at 47.2 percent (see Figure 18). In 2010, "too expensive" became the second most-cited reason despite slipping 0.6 percentage points, to 18.0 percent, during 2009-2010. The third most prominent reason, "no/inadequate computer," declined to 15.7 percent during the same period.

The second sub-category, households that lacked an Internet connection, but had users outside the home, accounted for 28.9 percent of the "no home broadband" category and included different reasons for non-adoption. For this group, cost ranked highest at 38.7 percent in 2010 (see Figure 19). "Don't need/not interested" increased by almost six (5.9) percentage points from 2009 to 22.6 percent in 2010. "No/inadequate computer" and "can use elsewhere" both declined over the two-year period, settling at 14.6 and 13.7 percent, respectively. Lack of availability decreased one percentage point to 1.7 percent in 2010.

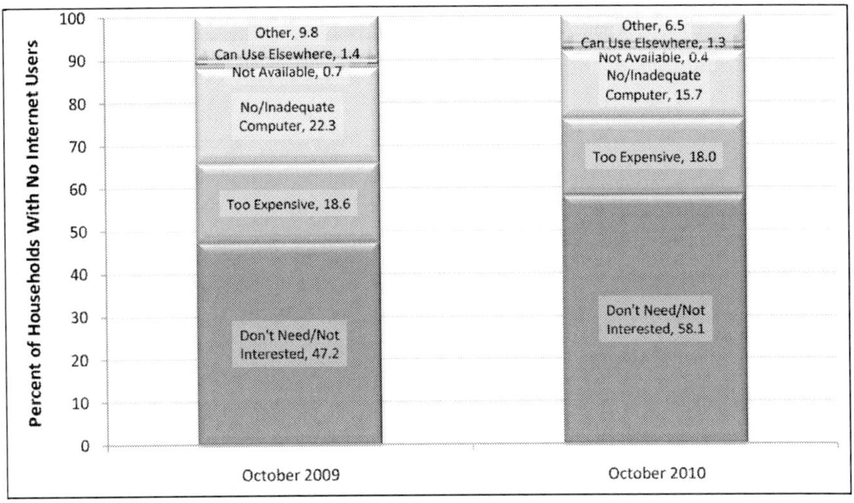

Figure 18. Main Reason Provided by Non-Internet Users, 2009-2010.

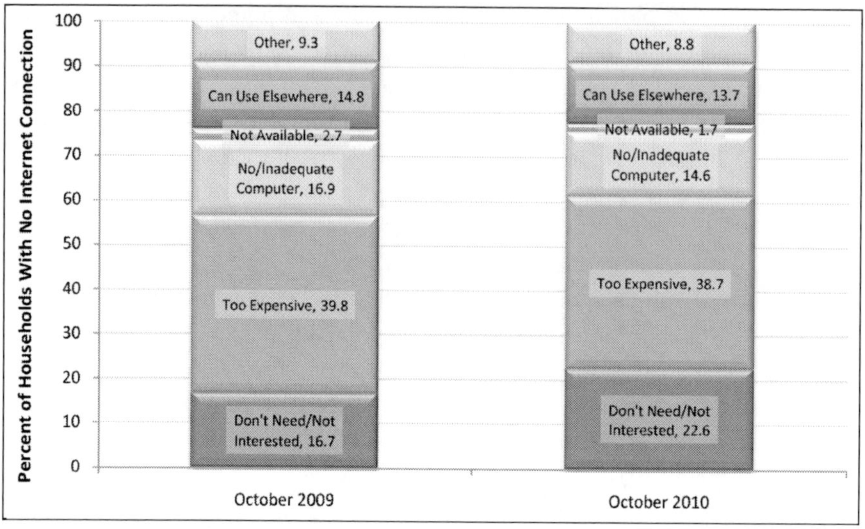

Figure 19. Main Reason for No Internet Connection at Home, 2009-2010.

The third sub-category, households with only dial-up Internet access, was the smallest group at 8.9 percent of households lacking broadband. As with the second sub-category ("only Internet use outside the home"), "too expensive" was the most frequently cited main reason for non- adoption, though dropping markedly from 2009 (41.3%) to 2010 (33.5%) for the dial-up group (see Figure 20). By 2010, "don't need/not interested" had increased almost five percentage points, nearly overtaking cost for the top spot at 32.5 percent. Lack of broadband availability also ranked highly for this group, rising almost seven percentage points between 2009 and 2010. This surprising development might be explained by the difficulty of overcoming lack of availability as a barrier, as compared with other reasons, as the pool of dial-up users shrinks rapidly (from 5.6 million households in 2009 to 3.4 million in 2010).

CONCLUSION

Today, more than two-thirds (68.2%) of U.S. households enjoy broadband Internet access at home. This represents an increase of almost five percentage points since 2009 (63.5%) and more than 17 percentage points since 2007. As a group, an estimated 209 million Americans about 72 percent of all adults and children ages three years and older use the Internet somewhere, whether it be at home, the workplace, schools, libraries, or a neighbor's house. Americans have developed a strong preference for high-speed connections and the applications they enable. Why does it matter? Available evidence supports the notion that broadband can create significant economic benefits for those who adopt it. In a global economy and the huge market it represents, this advantage could make the difference between those who succeed and those who do not. This resource can impact our nation's job base, productivity, competitiveness, economic growth, and ultimately, our standard of living.

Given the high stakes, it matters that we are leaving some individuals behind, particularly with respect to high-speed Internet at home. The good news is that the digital divide has receded somewhat in recent years. Non-adoption rates in the U.S. for home broadband (31.8% of households), home Internet (28.9% of households), and Internet used anywhere (28.3% of all persons) comprised less than one-third of the nation and were at new lows. Over 80 million (81.6 million) households have adopted broadband, but 38.0 million do not have such connectivity. Unfortunately, too many people with low incomes, less education, or disabilities, as well as unemployed individuals or seniors, certain minorities, and non-family households, are on the wrong side of the divide. The rural-urban disparity has decreased, but still persists. Major differences still exist across states.

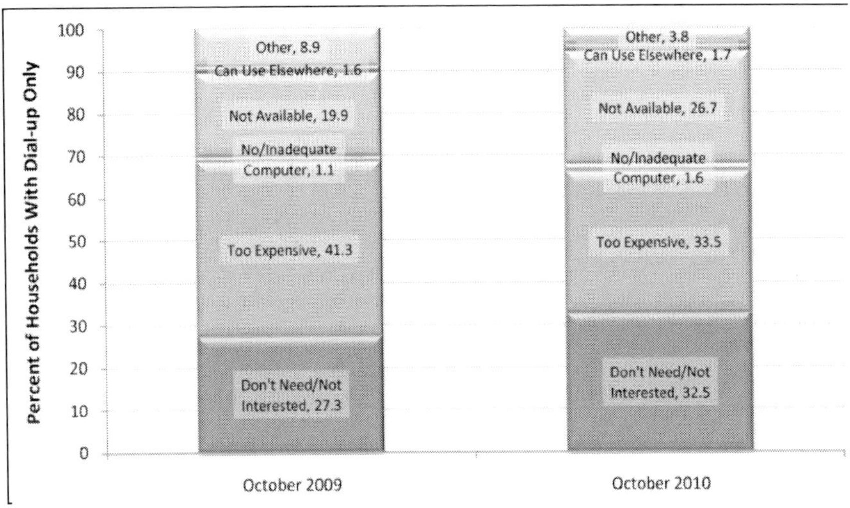

Figure 20. Main Reason for No Broadband in Households with Dial-up Internet Access, 2009-2010.

Overall, the two most frequently cited reasons for not having broadband Internet access at home were “don’t need/not interested” in it (46%) or “too expensive” (25%). “No/inadequate computer” (14%) ranked third. This ranking order held for the main sub-category of this group (non-Internet users), but the cost factor rated first for dial-up households and those that only used the Internet outside the home. Cost also became more important in such decisions for lower- income, and Black non-Hispanic or Hispanic households. Urban and rural households both ranked “don’t need,” “too expensive,” and “no/inadequate computer” as their top three reasons, but lack of broadband availability was much more important for rural areas (9.4%) than urban locales (1.0%).

This study serves to identify demographic and geographic trends, particularly for home broadband adoption in the United States, and to understand the main reasons why some U.S. households remain unconnected. Coupled with the NTIA-FCC interactive National Broadband Map, also issued today, these facts can be the linchpin for sound public decision-making and ensure that all Americans participate in the twenty-first century economy. Finally, we have made these findings and the associated data available for public use, including use by researchers and policy analysts, to further the policy debate and, ultimately, to enhance the public’s well-being.

End Notes

[1] This analysis stems from the 2010 Current Population Survey (CPS), which found that a household with at least one of the following Internet access services had broadband: DSL, cable modem, fiber optics, mobile broadband plan for a computer or a cell phone, satellite, or “some other service.”

[2] In the United States, broadband diffusion achieved 50 percent penetration within approximately eight years, while the other technologies took much longer to reach the same threshold: VCRs (12 years), cell phones (14), cable (15), color TVs (18), personal computers (19), radios (28), electricity (52), and telephones (71). *See* Adam Thierer, *On Measuring Technology Diffusion Rates*, The Technology Liberation Front (May 8, 2009), *http://techliberation.com/2009/05/28/on-measuring-technology*; Economic Statistics Administration & National Telecommunications & Information Administration, A Nation Online: Entering The Broadband Age, at fig.2 (2004); *infra*, at 7 fig.1.

[3] Leonard Waverman, London Business School, & LECG, *Connectivity Scorecard* (2009), *available at* http://www.nokiasiemensnetworks.com/news-events/press-room/press-releases/study-shows-significant-economic-benefits-from-broadband-if-ov.

[4] *See* John Deighton et al., *Economic Value of the Advertising-Supported Internet Ecosystem* (2010), *available at* http://www.iab.net/media/file/Economic-Value-Report.pdf.

[5] Sharon Gillett et al., *Measuring the Economic Impact of Broadband Deployment: Final Report to the Economic Development Administration* (2006), *available at* http://www.eda.gov/PDF/2006%20Measuring%20Broadband%20Report.pdf.

[6] For household-level estimates based on the total sample, the error attributable to sampling and other random effects at the 90 percent confidence level is no more than plus or minus 0.35 percentage points based on a standard error (SE) of 0.21 percentage points. For results based on Internet households, the margin of sampling error is no more than plus or minus 0.41 percentage points, based on a SE of 0.25 percentage points.

[7] NTIA has wholly or partially funded the surveys, with the help of the Economics and Statistics Administration, for the CPS Computer and Internet Use Supplements in 1994, 1997, 1998, 2000, 2001, and 2003, and Internet Use Supplements in 2007, 2009, and 2010. The data became the basis for the Commerce reports, Falling Through The Net (1995, 1998, 1999, 2000) and A Nation Online (2002, 2004), and provided input into the NTIA report Networked Nation: Broadband In America (2007). October 2009 CPS data underpinned the first two reports in the Digital Nation report series, issued in February and November 2010. website at http://www.ntia.doc.gov.

[8] For historical CPS data used in Data this " at report, see http://www.bls.census.gov/cps_ftp.html #cpssupps.

[9] *See* National Telecommunications & Information Administration, Digital Nation: 21st Century America' S Progress Toward Universal Broadband Internet Access, at fig. 2 (Feb. 2010), *available at http://www.ntia.doc.gov/reports/2010/NTIA_ internet_use_report_Feb2010.pdf.*

[10] This figure statistically draws from a pool of adults 25 years and older, thereby excluding a skewing of the results that would arise from including younger persons, such as current high school students.

[11] While not reflected here, significant disparities can exist among the sub-groups that make up a racial or ethnic category.

[12] *See* Economic Statistics Administration & National Telecommunications & Information Administration, Exploring The Digital Nation: Home Broadband Internet Adoption In The United States, at ch.7 (Nov. 2010), *available at* http://www.ntia.doc.gov/reports/2010/ESA_NTIA_US_Broadband_Adoption_Report_1108 2010.pdf.

[13] As set forth in Exploring The Digital Nation, *supra* note 12, when controlling for factors such as income, education, age, race, ethnicity, foreign-born status, household size, and geographic location, the adoption gap between those with no disability and those with a disability equaled only five percentage points in October 2009. This report does not undertake that "marginal effects" analysis.

[14] Because a household measure is less useful than a per person measure in gauging Internet or broadband use outside the home, this sampling does not use household measures. The universe of persons in this section includes all Americans three years old and above.

[15] The exception occurred in the September 2001 *See* Economic Statistics Administration & National Telecommunications & Information Administration, A Nation Online: How Americans are Expanding Their Use of the Internet, at fig 8-3 (2002).

In: Digital Nation: Broadband Internet Adoption... ISBN: 978-1-61324-556-9
Editors: Thomas N. Eaton et al.

Chapter 3

BROADBAND ADOPTION AND USE IN AMERICA: OBI WORKING PAPER SERIES NO. 1

John B. Horrigan

SUMMARY OF FINDINGS

The Federal Communications Commission's October- November 2009 survey finds that nearly two-thirds (65 percent) of American adults use high-speed Internet connections to go online from home.

The FCC conducted a survey of 5,005 Americans in October and November 2009 in an effort to understand the state of broadband adoption and use, as well as barriers facing those who do not have broadband at home. The main findings are:

- 78 percent of adults are Internet users, whether that means broadband, dial-up, access from home or access from someplace other than home.
- 74 percent of adults have access at home.
- 67 percent of U.S. households contain a broadband user who accesses the service at home.
- 65 percent of adults are broadband adopters. The discrepancy of two percentage points between household and individual home use is

because some survey respondents are nonbroadband users but live with someone who, at home, is.

- 6 percent of Americans use dial-up Internet connections as their main form of home access.
- 6 percent are Internet users but do not use it from home; they access the Internet from places such as work, the library or community centers.

For the purposes of this report, home broadband users are those who said they used any one of the following technologies to access the internet from home: cable modem, a DSL-enabled phone line, fixed wireless, satellite, a mobile broadband wireless connection for your computer or cell phone, fiber optic, T-1. In other words, home broadband users opt in to that classification through a survey question not by adhering to definition of broadband by speed that might be read to them.

The main dividing lines for access are along socioeconomic dimensions such as income and education.

- 46 percent of adults whose highest level of education is a high school degree are broadband users at home; 82 percent of adults who have attended or graduated from college are broadband users at home.
- 52 percent of Americans in households with annual incomes of $50,000 or below have broadband at home, compared with 87 percent of those in households with incomes above that level.
- Among low-income Americans–those whose annual household incomes fall below $20,000–broadband adoption stands at 40 percent.

African-Americans and Hispanics trail the average in broadband access, although gaps have narrowed since early 2009.

- 59 percent of African-Americans have broadband at home.
- 49 percent of Hispanics (English and Spanish speaking) have broadband at home.
    - For Hispanics who took the survey in Spanish, broadband adoption is only 20 percent.
    - For Hispanics who opted to take the survey in English, 65 percent have broadband.

These figures represent increases from levels registered in surveys conducted in early 2009 by the Pew Research Center, which found in April that 46% of African Americans and 40% of Hispanics (English and Spanish speaking) used broadband at home.

Some 42 percent of Americans with disabilities have broadband at home.

The FCC survey asked adults a series of six questions to determine whether a respondent should be classified as having a disability. Nearly one-quarter (24 percent) responded "yes" to at least one of the questions, indicating they have a disability; their broadband adoption rate is two-thirds the national average. Looking at the data differently, 39 percent of all Americans without broadband have some type of disability.

Senior citizens (those over the age of 65) continue to trail the national average in broadband adoption with a 35 percent broadband-at-home penetration rate.

- Nearly half (48 percent) of senior citizens are Internet users, regardless of connection type.

On average, Americans pay nearly $41 per month for broadband service, but half of those who receive their broadband in a bundle with other services cannot identify the Internet portion of their bill.

- When asked the level of their monthly Internet bill, to the nearest dollar, broadband users, on average, cited a figure of $40.68
- Most users (70 percent) receive broadband bundled with other services. Of these, 50 percent could specify their bundled monthly price: $37.70.
- Among broadband users who subscribe to a stand-alone high-speed service at home (i.e., they do not have service in a bundle), the average monthly bill reported was $46.25.

Broadband users overwhelmingly view the social aspects of the Internet as very important to them, while watching TV, videos or movies online and playing games were deemed less important.

When asked what types of online applications are important to them, here is what broadband users said were *very important* to them.

- 68 percent cited the Internet's capacity to ease communication with family and friends.
- 39 percent said the ability to use the Internet to keep up with community news.
- 35 percent mentioned the ability to share content, such as photos, videos or text.
- 24 percent cited online shopping.
- 10 percent said watching television shows, videos or movies online.
- 9 percent identified playing games online is very important.

Although most users of broadband at home have positive perspectives on the Internet, worries about inappropriate online content and security of personal information can put a damper on online activity.

Most broadband users see the Internet as a tool for learning and productivity.

- 81 percent strongly agree that the Internet is a valuable resource for information and learning.
- 74 percent strongly agree that it is important for children to learn to use the Internet.
- 56 percent strongly agree that people can be more productive using the Internet.

Nonetheless, broadband users view the Internet as having pitfalls.

- 56 percent strongly agree that too much pornography and offensive material is online.
- 39 percent strongly agree that it is too easy for personal information to be stolen online.
- 24 percent strongly agree that the Internet is too dangerous for children.

Users with positive outlooks about the Internet tend to be more active online (as measured by how many online activities they do) than those with less upbeat views on the Internet. Similarly, broadband users with strong concerns about the three pitfalls cited are *less* active online than those who do not share those views. These relationships do not show cause and effect, but they do suggest how attitudes influence the way people engage with the Internet.

Non-adopters are almost 50 percent more likely than broadband users to say they believe it is too easy for personal information to be stolen online.

- By a 57 percent to 39 percent margin, non-adopters of broadband at home say they strongly agree that it is too easy to have their personal information stolen online.

This is one factor linked to their lower likelihood of adoption. Although this concern is not necessarily a causal factor behind non-adoption decisions, there is a significant positive correlation between high levels of worries about personal privacy and non-adoption.

Broadband users exhibit varying degrees of understanding of digital concepts. That, in turn, influences what they do online.

Respondents received a series of questions asking how well they understood various terms relating to computers and the Internet; such questions serve as proxy measures for people's online skill levels. What follows is the share of broadband users who say they understood *very well* the listed terms:

- 61 percent–refresh or reload
- 44 percent–operating system
- 42 percent–Internet browser cookie
- 41 percent–JPEG file
- 40 percent–spyware or malware
- 16 percent–widget

Those with greater understanding of these terms are heavier online users. Some 29 percent of broadband users said they did not understand *any* of the listed terms very well, while 24 percent understood five or six of the terms very well. On average, the former group engaged in about half as many online activities as their more informed counterparts. Again, a causal relationship is not suggested here, but the findings indicate how skills can shape how heavily they use the internet.

People take advantage of multiple devices and services to go online, but these are usually supplementary access paths for them.

- 30 percent of American adults have used a handheld device (e.g., cell phone or smart phone) to access the Internet.[1]
- This behavior is more prevalent among minority groups.
  - 39 percent of African-Americans have accessed the Internet with a handheld device.
  - 39 percent of Hispanics have accessed the Internet with a handheld device.
- 84 percent of those who use the Internet via a handheld device have broadband at home. That figure is somewhat lower for African-Americans (78 percent) and Hispanics (68 percent).
- 15 percent of all Americans use a mobile wireless broadband service with their laptop computers.[2]
  - The vast majority (92 percent) of wireless broadband users have wireline broadband at home.
- Broadband users also access the Internet from places other than home.
  - 67 percent have at some time used the Internet at the homes of friends or family.
  - 62 percent have used the Internet at work.
  - 33 percent have used the Internet at a public library.
  - 30 percent have used the Internet at school.
  - 13 percent have used the Internet at a community center.
  - 5 percent have used the Internet at a place of worship.

Thirty-five percent of Americans do not use broadband at home. They fall into three categories, each with distinct demographic characteristics.

- 22 percent of adults are not Internet users. They are the oldest non-adopting group, with a median age of 60, and include the highest share of Hispanics (at 20 percent). Some 84 percent have high school degrees or less and half live in households with annual incomes of $30,000 per year or less.

- 6 percent of adults have dial-up connections at home. Nearly one-third (29 percent) of them live in rural areas, twice the rate for broadband users, and one-quarter (23 percent) have college degrees, which modestly lags the national average. The median age is 53.
- 6 percent of adults are Internet users, but do not have access from home. The group is relatively young (the median age is 38) and female (59 percent), but they tend to have low incomes and low levels of educational attainment.[3]

Of these non-adopters, 12 percent say they cannot get broadband where they live. This translates into a 4 percent share of Americans–on the basis of their reports on infrastructure availability in their neighborhood–who say they are unable to obtain broadband because it is not available. This means that 31 percent of all Americans can get service but do not.

There are three primary reasons why the 35 percent of non-adopting Americans do not have broadband: cost, lack of digital literacy and broadband is not sufficiently relevant for them to purchase it:

- 36 percent of non-adopters cite **cost** as the *main* reason they do not have high-speed Internet at home. This breaks out in the following ways:
    - 15 percent say the price of the monthly bill is too much for them.
    - 10 percent say the cost of a computer is too much.
    - 9 percent say they do not want a long-term service contract or cannot afford the installation fee.
    - 2 percent cite a combination of these reasons.
- 22 percent of non-adopters cite factors pointing to lack of digital literacy as the main reason they are not online. These include people who are not comfortable with computers or, for non-internet users, are "worried about all the bad things that can happen if I use the Internet." As people who cite digital literacy as barrier tend to be older (the median age is 62), concerns about the safety of the online environment is understandable.
- 19 percent of non-adopters do not have broadband because they question its relevance to their lives. They do not believe digital content is sufficiently compelling to justify getting it. Specifically, these non-adopters say the Internet is a "waste of time," do not think there is anything worth seeing online and (for dial-up users) say they

are content with their current service. Dial-up users make up a disproportionate share of those citing lack of relevance as a barrier.

Non-adopters concerned with cost would be willing to pay, on average, $25 per month for broadband.

Non-adopters who cited the monthly cost of broadband as a reason they did not have service received a follow-up question asking them to estimate how much they would pay for service. Among this group:

- 52 percent were able to provide an estimate; it averaged $25 per month.
- 28 percent answered "don't know" to this question.
- 20 percent said they were not willing to pay anything for broadband.

Some 65 percent of those who estimated their willingness to pay (WTP) cited a figure of $20 per month or more.

- If all of those who cited this figure had a service offering at $20 before them *and* took service, broadband adoption would be 6 percentage points higher in the United States.

Nearly all (91 percent) who provided a WTP estimate identified a figure of $10 per month or more.

- If all of those who cited this figure had a service offering at $10 before them *and* took service, broadband adoption would be 8 percentage points higher in the United States.

Non-adopters have a variety of outlooks on the Internet and information and communications technologies (ICTs).

Some non-adopters, notwithstanding worries they may have about the Internet and their capacity to use it, have a hopeful outlook on the benefits it may confer. Many are users of ICTs, just not broadband at home. Among nonbroadband adopters:

- 80 percent have premium television, i.e., either satellite or cable.
- 70 percent have cell phones.

- 42 percent have at least one working computer at home.

With respect to attitudes, most express high degrees of worry about the privacy of their personal information and inappropriate online content:

- 65 percent strongly agree there is too much pornography and offensive material on the internet.
- 57 percent strongly agree that it too easy for their personal information to be stolen online.
- 46 percent strongly agree that the internet is too dangerous for children.

At the same time, many non-adopters see the upsides to online access

- 59 percent strongly agree that the Internet is a valuable tool for learning.
- 54 percent strongly agree that it is important for children to learn to use the Internet.
- 37 percent strongly agree that people can be more productive using the Internet.

One-quarter (24 percent) of non-adopters have had experience with broadband, meaning they once had service at home or have used it at work or someplace else:

- 17 percent of all non-Internet users had home access in the past.
    - Of this group, 49 percent said they had home highspeed Internet access.
- 46 percent of dial-up or "not-at-home" Internet users have used a broadband connection from somewhere other than home, such as at work, school or a friend or family member's house.

Together, users in these two groups with direct experience with broadband equate to 24 percent of the non-adopting population, or 8 percent of all adults.

"Proxy Internet" use is evident for 22 percent of non-Internet or "not-at-home" users who live with someone with online access. Among these users:

- 16 percent ask the Internet user in the home to carry out an online task at least once a week.
- 20 percent ask the Internet user in the house to carry out an online task about once a month.
- About half of this "proxy access" is done using a broadband connection in the home.

When it comes to outlooks toward the Internet and levels of ownership of ICT products, non-adopting Americans fall into four categories, each with different barriers to broadband adoption.

Digitally Distant non-adopters make up 10 percent of the general population, and members of this group do not see the point of being online. Few in this group see the Internet as a tool for learning and most see it as a dangerous place for children. This is an older group (the median age is 63), nearly half are retired and half say that either *lack of relevance* or *digital literacy* are barriers to adoption.

The **Digital Hopefuls** make up 8 percent of the population. They like the idea of being online but lack the resources for access. Few have a computer and, among those who use one, few feel comfortable with the technology. They are most likely to cite *cost* as a barrier to adoption, with affordability of the computer playing an important role. They are also more likely than average to say *digital literacy* is a barrier. Demographically, this group is heavily Hispanic (26 percent), has a high share of African-Americans (20 percent) and is low-income.

The **Digitally uncomfortable** make up 7 percent of the population, and are the mirror image of the Digital Hopefuls; they have the resources for access but not a bright outlook on what it means to be online. Nearly all of the Digitally Uncomfortable have computers, but they lack the skills to use them and have tepid attitudes toward the Internet. This group reports a *variety of barriers*, including cost, lack of available infrastructure where they live, low perceptions of the Internet's relevance and low digital literacy.

The **Near Converts**, who make up 10 percent of the population, have many of the same characteristics of broadband adopters. They have high rates of computer ownership, positive attitudes about the Internet and are, relative to other non- adopters, youthful (at a median age of 45). Many are dial-up or "not-at-home" users, and *monthly access cost* is the largest reason for non-adoption among this group.

Looking across various population segments reveals differences in what they do online and what keeps them from having broadband.

Families: Parents with minor children at home are more likely than average to have broadband at home, by a 75 percent to 65 percent margin. The strong majority (87 percent) has at least one working computer at home, leaving a 13-percentagepoint gap between adopters and those with a functioning computer. Among those without broadband, nearly half (48 percent) cite cost as the main barrier to adoption, with 24 percent specifically pointing to the level of the monthly access fee.

Low-income Americans: Some 40 percent of low-income Americans (with annual household incomes at $20,000 or below) have broadband, compared with 91 percent among those living in homes with annual incomes above $75,000. Low-income broadband users are more likely than well-off broadband users to look for or apply for a job online – by a 77 percent to 60 percent margin. Not surprisingly, 47 percent of this group identifies cost as the biggest barrier to broadband adoption. That is twice the rate of the (relatively few) upper-income nonbroadband adopters.

African-Americans: Although African-Americans trail the average in overall broadband adoption, the adoption gap does not exist for adults under age 30; some 75 percent of African-Americans in this younger age group have broadband, which equals the average. By contrast, 21 percent of African-American senior citizens have broadband; that compares with the 35 percent average. African-American broadband users are highly likely to have used their broadband connection to look for or apply for a job; 83 percent versus the 60 percent average. African-Americans with broadband are also more likely to take a class for credit online (37 percent versus the 26 percent average). With respect to barriers to adoption, 42 percent of African-Americans say cost is the main reason they do not have broadband.

Hispanics: Half (49 percent) of Hispanics have broadband at home. While young African-Americans have broadband access on par with the average, "under 30" Hispanics trail the average for their age cohort, 57 percent versus 75 percent. Those Hispanics with broadband use it for downloading and streaming music (69 percent have done this, in comparison with the 52 percent average). They also find broadband very important for keeping up with news about their community; 52 percent say this compared with the 39 percent average. As to adoption barriers, 52 percent of Hispanic non-adopters cite cost, equally split between those who point to the monthly fee and those who say they cannot afford a computer.

People with disabilities: Of the 24 percent of respondents who have some sort of disability, 42 percent have broadband. The online activities of broadband-using people with disabilities are narrower in scope than others; that is, they do fewer things online. That may reflect difficulties some people with disabilities have in using the devices to get online or interacting with Web pages. Barriers for non-adopting people with disabilities do not differ significantly from the average.

Rural Americans: Fifty percent of rural residents have broadband, a rate that reflects in part the older and less wealthy rural population but also the lack of available infrastructure. One in 10 rural non-adopters say they cannot get broadband where they live. That is more than twice the average. Rural Americans *with* broadband, meanwhile, are as active as their urban and suburban counterparts in using the Internet for shopping and taking classes online, suggesting that they use broadband as a way to virtually access the benefits associated with urban or suburban living.

INTRODUCTION

The Federal Communications Commission in the fall of 2009 fielded a national survey of Americans under authority granted by the Broadband Data Improvement Act (BDIA). The BDIA, which became law in October 2008 but needed and received funding approval under the American Recovery and Reinvestment Act (ARRA), requires the FCC to conduct periodic surveys that explore the broadband experience of American consumers. Specifically, the BDIA directs the survey to probe the types of technology people use for their home broadband service, what they pay, the applications they use and, for those who do not use broadband, the barriers to adopting broadband.

This paper reports the national findings from the random digit dialing telephone survey that interviewed 5,005 adult Americans, 2,671 of whom are broadband-at-home users. The remaining 2,334 either do not have broadband at home, report that they do not use the Internet or say they are Internet users but without access at home. The number of non-adopters surveyed represents an oversample of this population; this was done to facilitate statistical analysis of non-adopters.

The Methodology section in the report's appendix provides detail on the sampling design for this survey, and also discusses such issues as response rate and use of cell phone numbers in the sample. Cell phone numbers were used in the sample in recognition of the fact many Americans do not have landline telephone service at home, but rather rely only on their cell phones. Overall, interviewers conducted 31% of the surveys with respondents who were on a cell phone; 13% said they were "cell phone only" households, while 18% had a landline, yet were contacted on their cell phone.

It is also worth noting that the data presented throughout is weighted to take into account characteristics of the sample, including the oversample of non-adopters. The weighting corrects for the fact that, in the figures noted in above, 2,671 adopters does not represent 65% of 5,005 respondents, the home broadband adoption figure highlighted at the outset. In results reported within, percentage figures represent weighted results; the number of cases reported for various sub-samples are actual number of cases.

For the survey, the margin of error based on results based on the entire sample of 5,005 is plus or minus 1.6 percentage points. For results on 2,671 home broadband adopters, the margin of error is plus or minus 2.1 percentage points; for the oversample of non-adopters (n=2,334) the margin of error is plus or minus 2.2 percentage points. Finally, results based on all internet users (n=3,555), the margin of error is plus or minus 1.8 percentage points. Interviewers conducting the survey provided a Spanish-language option for respondents wishing to take the survey in Spanish.

The FCC's survey results compare favorably to a similar survey conducted in the fall of 2009 by the Census Bureau for the National Telecommunications and Information Administration, which found that 63.5 percent of American households used high-speed internet to go online.[4]

What is New about this Survey

The survey is unique for the FCC, in that it is the first time that the Commission has conducted its own survey of end-users of the Internet. The Commission, using data gathered from FCC Form 477 (a semi-annual reporting document for telecommunications entities) has previously reported on the number of broadband lines in service to end users based on carrier-reported data at the Census tract level.[5] This survey, as it involves interviewing adults, presents adoption data based on what users say about how they access the Internet at home.

The survey is distinct in a broader sense owing to its focus on non-adopters of broadband at home. There have been other national surveys that have examined why people do not use the Internet, most notably the Pew Internet & American Life Project's 2002 survey on why people did not use the (mainly) dial-up Internet.[6] Other Pew Internet research has analyzed non-adoption of broadband based on a few questions in national surveys.[7]

Finally, the FCC survey develops segments of non-adoption. This segmentation permits non-adopters to be placed in categories that yield insights into the reasons people do not have broadband–and possible solutions that might be targeted at non-adopters. The British regulator Ofcom undertook a similar segmentation of non-adoption[8] but this is the first time national data collected by the U.S. government has been examined in that way.

I. Adoption: Who Uses Broadband at Home–and Who Doesn't

Who Uses Broadband at Home–and Who Doesn't

The October-November FCC survey finds that a strong majority of Americans use the Internet. Among adults, 78 percent are Internet users, meaning they have access from home or someplace other from home, and with any type of connection (i.e., dial-up or broadband). The findings about access:

- 65 percent of adults use broadband at home.
- 67 percent of households have broadband. The survey asked non-adopters whether another person in the house uses the Internet with a home high-speed connection. Affirmative answers to that question add two additional percentage points to overall penetration.
- 6 percent of Americans have dial-up Internet connections at home.
- 6 percent of Americans are Internet users but do not access the Internet from home.

This means that 22 percent of Americans are not Internet users (1 percent of respondents did not know the type of Internet connection they have at home and thus are not classified as broadband, dial-up or "not at home" online users).

Although broadband access and reasons for not having it are the principle themes of the survey, respondents were asked about several other assets pertaining to information and communications technology (ICT), such as use of computers, TVs and cell phones. Specifically:

- 86 percent of Americans have a cell phone.
- 65 percent have cable TV at home.
- 66 percent have a desktop computer at home.
- 52 percent have a laptop computer at home.
- 29 percent have satellite TV.

Combining cable and satellite TV, as well as desktop and laptop adoption, yields the following figures for home computer and home premium TV use:

- 86 percent of Americans have either cable TV or satellite TV.
- 79 percent of Americans have either a desktop or laptop computer at home.

The appendix to this paper provides detail on adoption patterns for each of these technologies across demographic and socio-economic categories.

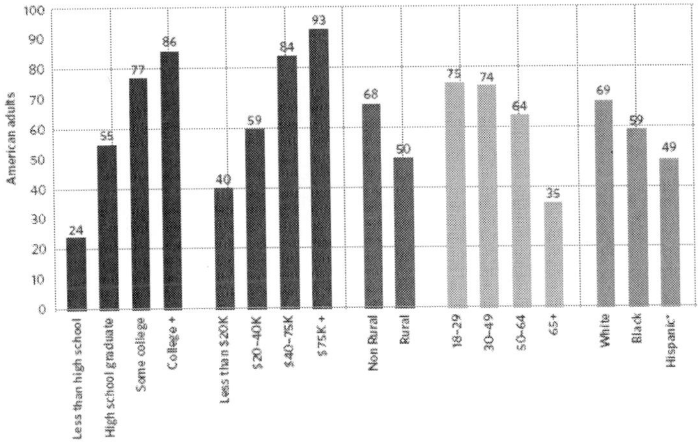

*Hispanics includes both English and Spanish-speaking Hispanics

Exhibit 1. Broadband Adoption by American Adults by Socio-Economic and Demographic Factors.

Dividing Lines on Broadband Access

Broadband adoption is not distributed evenly in the population *(See Exhibit 1)*. The differences in adoption break out prominently along two particular dimensions, which are not mutually exclusive:

Education: Among adults, roughly half of Americans have had some college experience (even if they have not or did not graduate). Among those who have some college experience, 82 percent have broadband at home, compared with 46 percent of those whose highest level of educational attainment is a high school degree.

Income: Americans in the lower half of the income distribution demographic, meaning an annual household income of $50,000 per year or less, are much less likely to have broadband at home than those with higher incomes. Among respondents reporting household incomes of $50,000 or less, 52 percent have broadband at home, far below the 87 percent adoption rate for those above that income threshold.

Adoption Patterns for Key Population Segments

Surveys conducted by the Pew Research Center's Internet & American Life Project found that broadband adoption within two minority groups, African-Americans and Hispanics, lags the average and, for African-Americans, has grown very little from 2007 to the beginning of 2009. The FCC survey shows some breaks from those patterns.

African-Americans: FCC figures found 59 percent of African-Americans had broadband at home in the October- November survey, up considerably from 46 percent in Pew's April survey.

Hispanics: Slightly less than half (49 percent) of Hispanics have broadband at home in the FCC survey compared with 40 percent of Hispanics in a Pew Research Center April 2009 survey. In the FCC survey, 3 percent of respondents chose to take the survey in Spanish; among these Spanish speakers, broadband adoption was 20 percent.

People with disabilities: The FCC survey finds that 42 percent of those who identify themselves with disabilities has broadband at home. A July 2003 survey from the Pew Internet Project showed that 10 percent of those with disabilities had broadband at a time when 18 percent of Americans had broadband.[9]

The FCC survey sampled all Americans, but for several racial or ethnic categories, there were not enough respondents to draw statistically reliable inferences. For Asian-Americans, American Indians, and Alaskan natives, the sample yielded fewer than 100 respondents in each group. The first two groups in particular have a sizable population that may not speak English or that have low telephone penetration rates. Because of that and the small sample of respondents, it is inadvisable to report results.

How People Connect to the Internet at Home

The survey asked respondents what type of Internet connection they use at home. The question, which is the same one used in surveys conducted by the Census, allowed respondents to pick any of the categories listed that the respondent believed described his home broadband connection. In other words, respondents could pick more than one type of home broadband connection from the list *(See Exhibit 2).*

The answers underscore the difficulty in asking people about the technological dimensions of their home Internet access points. Although it is possible that some consumers have more than one home high-speed connection, these figures suggest that most Americans have two broadband connections. It is likely that a significant number of consumers do not know the details of their home Internet connections. Some, for instance, may confuse a wireless home network working off wireline broadband with a "fixed wireless provider." Given the opportunity to pick more than one category, in conjunction with uncertainty over how they connect to the internet at home, some respondents inaccurately chose more than one category of connection type.

Notwithstanding the possible confusion reflected in the survey responses, it seems likely that the vast majority of home broadband access is wireline. In fact, estimates place wireless home broadband access at 2 percent of homes– that would include fixed wireless or satellite service.[10]

Exhibit 2. At Home, What do You Now Use to Connect to the Internet (Figures as a Percent of Home Internet Users)

Cable modem	58
DSL-enabled phone line	44
Mobile broadband wireless connection for your computer or cell phone	44
Fixed wireless provider	29
Dial-up telephone line	12
Satellite connection	10
A fiber optic connection such as FiOS	10
T-1 connection	5

Source: Federal Communications Commission survey of 5,005 adult Americans, October-November 2009.

The survey also made extensive efforts to determine whether a respondent had dial-up service at home or not. First, interviewers asked respondents first whether they connect at home "through a slow-speed connection such as dial-up or...a high-speed, broadband connection." Among home users (72 percent of respondents), 8 percent said they had a slow-speed connection.

Second, a follow-up question for the 8 percent who said they had a slow speed connection asked them to confirm that they "now use a dial-up connection to the Internet at home, not a higher-speed connection." Most (93 percent) who first said they had a slow-speed connection confirmed that they *now* have dial-up at home. For the purposes of this report, only those who were able to confirm in the second question that they have dial-up at home are defined as home dial-up users. This amounts to 6 percent of the entire adult population.

With the questioning approach isolating home dial-up users, home broadband users are defined, in this report, as respondents who chose any of the options listed in *Exhibit 2* that are home high-speed connections, that is, cable modem, a DSL-enabled phone line, fixed wireless, satellite, a mobile broadband wireless connection for your computer or cell phone, fiber optic, T-1. Thus, home broadband users opt in to that classification through a survey question not by adhering to definition of broadband by speed that might be read to them.

Monthly Cost of Service

As the BDIA directs, the survey also asked consumers what they pay per month for home broadband service. This question presents another challenge for respondents, separate and apart from knowing the details about home connection technology. Many consumers have Internet service bundled with other services (such as telephone or cable TV), making it difficult for the respondent to know what portion of the total bill reflects the cost of Internet service.

To address bundling, the survey asked consumers if they paid for various services (cell phone, landline phone, Internet, cable TV, satellite or wireless broadband) in conjunction with other services. If consumers said they paid for the Internet in a separate bill, they were asked to tell interviewers what their average monthly bill was to the nearest dollar. If respondents said their Internet service was bundled with another service, follow-up questions addressed what the monthly bills were for the entire bundle, whether the respondents knew the Internet portion and, if so, what the monthly Internet bill was.

This careful approach to bundling has the advantage of stripping out those who do not know the Internet portion of their bundle before they are asked to quantify their monthly Internet bill. Those who do not know are not asked to make an estimate. However, by appropriately excluding these respondents, the overall sample of respondents offering cost estimates is reduced. One-third of broadband users did not give an estimate of their monthly broadband payments.

For Internet users, bundling their bill with other services is common. A majority of home broadband users (70 percent) say their Internet service is bundled with another service in their monthly bill, with 29 percent saying they receive a stand-alone bill for broadband.

Among those who bundle Internet with another service, 50 percent said they knew what the Internet portion of the bill was, a figure that was the same for broadband and dial-up users alike.

The survey finds that:

- Overall, users report paying $40.68 per month for their broadband Internet connections.[11]
- Those who report they take service in a bundle (among those who can report what the Internet portion of the bundle is), the average bill is $37.70.

- Those with stand-alone broadband service report a monthly bill of \$46.25.
- Those use dial-up at home to go online report a monthly bill of \$22.98.

Several other sources gather data on what people pay for broadband and offer points of comparison. They are:

- The Pew Internet Project found in its April 2009 survey that users report an average monthly broadband bill of \$39.
- TNS Telecoms, through analysis of consumer bills, 90 percent of which are bundled offerings, finds an average broadband bill of \$34.50.
- Telogical Systems examines stand-alone nonpromotional offers of providers, and its figures show an average broadband bill of \$46 per month.
- Telogical Systems examines stand-alone nonpromotional offers of providers, and its figures show an average broadband bill of \$46 per month.

Telogical's figures for stand-alone prices track reasonably well with what respondents in the FCC survey say about stand-alone bills. Those in the FCC survey able to identify the Internet component of their bundle cited an average figure about \$3 more than the bill analysis of TNS Telecoms, which focuses mainly on bundled offerings.[12]

II. Online Behaviors: What Shapes Them and What Matters to Users

This section describes the kinds of online activities people engage in, as well as two factors that influence intensity of use. This involves analyzing: a) how attitudes toward the Internet influence how much they do online; and b) how the level of understanding of computers and the Internet influences what they do online. The section concludes by looking at what people find most important about their online activities.

Online Activities

The Internet, as has been well documented, is a means for communication, collaboration and content creation directed at users, shoppers, social networkers, information seekers and people searching for entertainment. The survey asked Internet users if they had ever engaged in particular online activities, with the topics chosen to reflect the variety of activities people may pursue online.

To economize on the survey length, half the respondents received eight of the listed activities in *Exhibit 3,* and the other half received the other eight. Among the first set of activities (designated with an asterisk in the chart), broadband users on average participated in five of them. For the second set, broadband users did 3.7 of them, on average.

Those averages are affected by a host of factors. Young people are generally more attuned to digital life, so it is no surprise that adults under the age of 30 are typically more active. Respondents in that group say, on average, that they do 5.9 and 4.6 of the online activities across the two sets. The better educated also do more things online: College graduates say they have done 5.3 and 3.9 activities, on average, of the activities listed. A table in the appendix details how online activities break out along demographic dimensions.

As influential as demographics may be in explaining the scope of what people do online, there are two other relevant factors: attitudes about the Internet (its strengths and hazards) and skills (using measures of people's understanding of computers and the Internet as a proxy for their level of skill).

Attitudes about the Internet

Respondents were read a series of six statements about the Internet–three about the Internet's upsides and three about its less attractive qualities *(See Exhibit 4).*

In general, broadband users who showed strong levels of concern about potential hazards online reported doing a narrower scope of online activities than respondents without such worries. Similarly, those with positive perspectives were more active online. It is important to note that worries about online dangers do not always translate into behavioral changes. People may register concerns about identity theft, but this may not prevent many users from sharing personal information at online social networking or shopping sites.

Exhibit 3. Online Activities of American Adults (Figures as a % of Users in Each Group)

	All Internet users	Dial-up users	Broadband users
Buy a product online*	78	56	83
Get local or community news*	75	55	80
Visit local, state or federal government Web site*	75	53	79
Use a social networking site*	52	41	55
Submit a review for a product or service*	52	36	55
Download or stream music*	47	22	52
Upload or share content*	45	26	48
Play games online*	46	38	48
Get international or national news	73	54	77
Bank online	63	43	69
Get information about or apply for a job	57	39	60
Get advice from a government agency about a health or safety issue	50	39	54
Download or stream video	38	18	42
Post to own blog or group blog	23	7	26
Take a class online	22	8	24
Play complicated role-playing games online	14	9	14

Source: Federal Communications Commission survey of 5,005 adult Americans, October-November 2009. Draft final results. For broadband users, n=1,378 for activities marked by * and n=1,278 for other activities. For dial-up user, n=212 for activities marked by * and n=247 for other activities.

The following analysis shows how scope of activity varies by respondents' attitudes. Recall that half the sample was asked about eight different online activities, and the other half asked about a separate set of eight activities. To show the interaction of attitude and activity, the analysis below reports the average number of activities (across both sets) among those who strongly agree with a specific attitudinal proposition. That is then contrasted, in percentage terms, with those who did not share that view.

The Internet is too dangerous for children: Those who strongly agreed with this did an average of 4.3 and 3.1 activities, 19 percent lower than those who do not share this view (averaged across both sets of activities).

It is important for children to know how to use the Internet: : Those who strongly agreed with this did an average of 5.2 and 3.8 activities, 16 percent more than those who do not share this view (averaged across both sets of activities).

The Internet is a valuable source for information and learning: As was the case for the importance of the Internet for learning among children, those who strongly agreed with this engaged in an average of 5.2 and 3.8 activities, 25 percent more than those who do not share this view (averaged across both sets of activities).

There is too much pornography and offensive material on the Internet: Those who strongly agreed did an average of 4.7 and 3.4 activities, 11 percent lower than those who do not share this view (averaged across both sets of activities).

People can be more productive using the Internet: Those who strongly agreed with this did an average of 5.2 and 4.0 activities, 18 percent more than those who do not share this view (across both sets of activities).

It is too easy for my personal information to be stolen online: Those who strongly agreed with this did an average of 4.6 and 3.4 activities, 11 percent lower than those who do not share this view (across both sets of activities).

Overall, broadband users who strongly registered any of the three worries listed above (theft of personal information, inappropriate content, online dangers for children) engaged in 4.8 and 3.5 of the online activities probed. The share of all broadband users who cited any of those three things as *strong* concerns is 70 percent and, when compared with the 30 percent of broadband users who share *none* of these concerns strongly, they do 14 percent fewer online activities. These comparisons do not mean that the attitudes in question *cause* or are sole drivers of the differences identified. However, they do indicate that how people perceive the Internet shapes how they use it. It is worth noting that the differences identified above are statistically significant when controlling for other respondent characteristics, such as age, race, education and income.[13]

Skills: Understanding Information and Communications Technology

The scope of activities people do online may also have to do with the skills they bring to the online experience. Measuring the level of skill or literacy for a user is challenging in a survey environment. Merely asking people to rate how well they can carry out a given task can be problematic, and research has found that such self-assessment about skill is not the best predictor of actual skill levels. A better predictor of actual skill comes from asking respondents to rate on a scale their understanding of various concepts relating to the Internet or computers.[14]

The FCC survey borrowed this approach, asking Internet users how well they understand six terms relating to the Internet or computers *(See Exhibit 5).*

Not surprisingly, education is strongly associated with the likelihood that people say they understand a particular concept very well *(See Exhibit 6).*

An important consequence of low levels of digital literacy is less engagement with online life. Those who say they understand the queried concepts "very well" are likely to do more of the online activities discussed. This correlation is significant when holding constant income, education, age and other factors that might also affect people's probability of doing a range of things online.[15]

Exhibit 4. Public Attitudes about the Internet (% of Broadband Users who Strongly Agree)

The Internet is a valuable source for information and learning	81
It is important for children to learn how to use the Internet	74
There is too much pornography and offensive material on the Internet	56
People can be more productive using the Internet	56
It is too easy for my personal information to be stolen online	39
The Internet is too dangerous for children	24

Source: Federal Communications Commission survey, October-November 2009, n=2,671 for broadband users.

Exhibits 7, 8 and 9 demonstrate this point by showing, for broadband users, how the number of online activities in which they engage increases in tandem with the number of computer and Internet terms they say they understand "very well." For the first set of activities, broadband users who did not say they understood any of the computer or Internet terms listed (29 percent of respondents) engaged in an average of 3.8 of the online activities about which they were asked. Those who understood very well all of them (24 percent of respondents) did, on average, 6.1 activities. The differences are more pronounced for the second set of activities, with those who understand all of the terms very well doing, on average, more than twice as many activities as those who understood none of them.

The Activities People Say are Most Important

What people do online and how they value a particular activity can differ for a number of reasons, often due to the context in which an activity is undertaken. More Internet users have, at some point, bought something online (78 percent) than have uploaded content they have created (45 percent) or participated in a blog (23 percent). However, seeing a video of a child's birthday party may be more important to some people than buying a product online – especially if offline shopping opportunities are abundant. Conversely, coming upon a hard-to-find good in the far-flung corner of the Web may be extremely important if the buyer is a collector or shopping for a special occasion.

Exhibit 5. Broadband Users' Understanding of Computer and Internet Concepts (Figures as a % of Users)

	Very well	Somewhat well	Not too well	Not well at all
Internet browser cookie	42	30	12	15
Spyware and malware	40	34	11	14
Operating system	44	29	11	16
Refresh or reload	61	22	5	11
Widget	16	13	15	54
JPEG file	41	22	8	27

Source: Federal Communications Commission survey, October-November 2009; n=2,671 for home broadband users.

Exhibit 6. By Education, Broadband Users' Understanding of Computer and Internet Concepts (% "Very Well" in Each Group)

	Less than High School	High school grad	Some college	College+
Internet browser cookie	23	32	44	50
Spyware and malware	26	33	41	47
Operating system	21	31	44	55
Refresh or reload	46	52	64	68
Widget	4	11	15	22
JPEG file	17	28	41	55

Source: Federal Communications Commission survey, October-November 2009; n=2,671 for home broadband users.

Exhibit 7. Number of Online Activities and Digital Literacy

	Average number of online activities (first set of eight activities)	Average number of online activities (second set of eight activities)
None	3.8	2.5
One term	4.9	3.3
Two terms	5.1	3.5
Three terms	5.4	4.2
Four terms	5.6	4.0
Five or more terms	6.1	4.8

Source: Federal Communications Commission survey, October-November 2009; n=1,378 for home broadband users for first set of activities and n=1,293 for home broadband users for second set of activities.

Even though context is key to interpreting how people value different online activities, the survey nonetheless sought to explore this by asking Internet users to assess the importance of six different categories of online activities. *Exhibits 8 and 9* show results for all users and broadband users when they were asked to assess how important each online activity was to them.

For most Internet users, the communicative aspects of the Internet are most important to them, followed by keeping up with news in the community and sharing content with others. Applications oriented to entertainment rate notably low, perhaps because (for watching TV shows or movies) excellent offline alternatives exist. Young broadband users are more likely to cite entertainment applications or online entertainment as "very important" than average, with 20 percent and 13 percent saying this. They are also more likely to say communicating with family and friends (79 percent) and sharing content (45 percent) are very important online activities.

Why New Users Get Online

What existing users value about online activities and what triggers new adoption can be two different things. To address this, the survey asked about new users' motives for getting online.

The level of Internet penetration has not grown substantially in recent years. As noted, the FCC survey finds that 78 percent of Americans use the Internet, whether that is at home or somewhere else, dial-up or broadband. Surveys conducted by the Pew Internet Project show that Internet penetration first exceeded 70 percent in early 2006, with that figure reaching 75 percent for the first time at the end of 2007. The FCC asked how long respondents have been using the Internet; just 6 percent report that they have been online for two or fewer years. Those who have been online for two or fewer years received follow-up questions asking them to cite the reasons they chose to get Internet access and then the *most* important reason for getting access.

Exhibit 8. Broadband Survey Users on the Importance of These Activities

% as a share of *broadband* Internet users at home	**Very important**	**Somewhat important**	**Not too important**	**Not important at all**
Making it easy to communicate with friends and family, even if they are far away	68	23	5	4
Keeping up with the news in my community	39	39	11	10
Sharing content with others, such as photos, videos or text	34	38	15	13
Shopping online	23	41	19	18
Watching TV shows, movies and other video online	10	26	24	39
Playing games online	9	18	23	50

Source: Federal Communications Commission survey, October-November 2009; n=2,671 for home broadband users.

Exhibit 9. Broadband Survey Users on Their Reasons for Getting Online (% Among Internet Users who Have Been Online for 2 Years or Less)

	% who cite reason	**% who cite as *most important* reason**
To e-mail and stay in touch with family and friends	61	31
To gain access to music, movies and other entertainment	46	9
To share my photos or videos with family and friends	41	3
My children wanted Internet access	41	7
My children needed it for school	35	19
I needed it for school	27	10
An Internet provider made a special offer too good to pass up	21	2
My job required online access	20	6

Source: Federal Communications Commission survey of adult Americans, October-November 2009. N=92 for new users.

As *Exhibit 9* shows, most new users cite social reasons as a motive for beginning to use the Internet, with e-mail communication and content sharing as chief reasons. Accessing video or other Internet entertainment online also was a factor for many new users. A great offer by an Internet service provider was not widely cited by new users. When pressed about the most important reason, social and educational uses are most prominent. E-mail is the main reason for one-third of new users, while school–either for themselves or children–is the main reason for another one-third of new users. It is worth pointing out the small number of cases (92) for the subset of new users who received these questions.

III. Alternative Access: Different Places, Different Platform

Access to the Internet is portable and shareable. People may take access with them or, if they cannot, they may access the Internet from different places. The survey looked at this in three contexts: handheld access, mobile broadband (from a laptop computer) and Internet access from places outside the home (often public access points).

The Places People Access the Internet

Although most online users have access from home, there are plenty of places outside the household where people can go online. To examine how people access the Internet away from home, respondents received questions that tried to fix whether they have gone online from someplace other than home and, if so, where. *Exhibit 10* shows the places from which people access the Internet.

These results show that 93 percent of Internet users have some form of access at home. Yet, it is apparent people access the Internet in other contexts as well–particularly at work or a friend or family member's house.

Exhibit 11 indicates that a majority of Americans (59 percent) who access the Internet do so in more than three places; 17 percent of Americans go online in 5 or more places.

Focusing on alternative access places that *exclude* home or work, 57 percent of all Americans (or 73 percent of Internet users) have at one point gone online at some place other than home or work, often at a friend or family member's house. Eliminating the latter from consideration, 39 percent of Americans (or 48 percent of Internet users) have gone online at a library, school, community center or church. It is worth noting that these figures do not include those who may access the internet at a local coffee shop.

Exhibit 10 . Broadband Survey Users on Where They Access the Internet

	All Internet users	Broad-band at home	Dial-up at home	Not-at-home users
Home	92%	100%	99%	7%
Friend or family member's house	65	67	48	67
Work	58	62	36	48
Public library	35	33	32	57
School	29	30	17	21
Community Center	14	13	9	25
Church	5	5	4	5
Number of cases	3,555	2,671	459	392

Source: Federal Communications Commission survey, October-November 2009.

Exhibit 11. Broadband Survey Users on the Number of Places They go Online

(% of Internet users)	
One access point	17
Two access points	24
Three access points	25
Four access points	17
Five or more access points	17%

Source: Federal Communications Commission survey, October-November 2009.

Exhibit 12 shows how incidence of using the internet at different access sites varies across demographic and socioeconomic characteristics of the respondent.

Use of the Handheld for Online Access and other Purposes

With Internet access increasingly in people's pockets, it is important to probe the activities people pursue using their handheld devices. Fully 86 percent of Americans have a cell phone. Estimates of the share of people with "smart phones" that permit online access vary, but Forrester Research places the share of the adult population with smart devices at 17 percent in November 2009.[16] The survey did not ask directly whether a respondent had a smart phone, but rather whether cell users did specific activities–some of which are likely to require having a smart phone.

Exhibit 13 shows the findings from these questions, with a breakout by age, as young adults are more frequent users of mobile devices than their elders.

Although basic communication functions dominate the portfolio of cell phone activities, there is a substantial incidence of information seeking, such as accessing Web pages or searching for directions. Twenty percent of cell phone users say they have downloaded an application to their device, suggesting that at least that share has a smart phone; the actual level of smart phone usage is probably higher since not all smart phone users are likely to download applications.

Exhibit 12. Use of Public Access Points and Respondent Characteristics

(% of those surveyed)	Public library	School	Community center	Church
Male	33	26	15	5
Female	38	31	13	5
Parents with minor children at home	35	28	13	5
Those who report they have a disability	35	22	16	4
18–29	56	58	26	5
30–49	33	26	13	4
50–64	25	14	8	4
65+	20	6	5	5
White (not Hispanic)	31	25	11	4
Black (not Hispanic)	51	36	21	6
Hispanic (English or Spanish speaking)	46	43	26	5
Less than high school	41	24	19	2
High school graduate	33	21	14	4
Some college	41	34	15	5
College +	31	31	11	6
Under $20K	55	41	25	4
$20K–$30K	44	30	20	6
$30K–$40K	41	33	13	5
$40K–$50K	30	23	12	4
$50K–$75K	28	24	10	5
$75K–$100K	31	27	11	5
Over $100K	28	30	10	4
Urban	38	33	16	5
Suburban	33	27	12	4
Rural	34	22	13	4

Source: Federal Communications Commission survey, October-November 2009.

As documented by the Pew Internet Project, use of mobile devices for non-voice data applications is particularly popular among African-Americans and Hispanics.[17] Unlike the Pew survey, the FCC survey had a Spanish-language option, which means data reported here on mobile use among Hispanics represents a more complete sample of the Hispanic population. Still, African-Americans and Hispanics, consistent with prior research, are the most frequent users of the handheld device for a variety of applications *(See Exhibit 14).*

Exhibit 13. Cell Phone Activity

	All adults	Age 18–29	Age 30–49	Age 50–64	Age 65+
% with a cell phone	86	94	93	86	66
% as a share of those with cell phone					
Send or receive text messages	66	94	79	51	15
Send or receive pictures	52	77	61	38	14
Send or receive e-mail	26	40	30	17	5
Send or receive Instant Messages	28	42	35	20	8
Access Web pages on the Internet	28	48	32	15	5
Get a map or directions to another location	27	44	30	16	8
Download an application to your cell phone	20	36	24	9	3
Download or stream music or video	17	32	18	8	3

Source: Federal Communications Commission survey of 5,005 adult Americans, October-November 2009.

Exhibit 14. By Race, Cell Phone Activities

	Whites	African-Americans	Hispanics
% with a cell phone	87	86	85
% as a share of those with cell phone			
Send or receive text messages	63	75	75
Send or receive pictures	50	61	58
Send or receive e-mail	24	34	28
Access Web pages on the Internet	24	36	36
Get a map or directions to another location	23	36	34
Send or receive Instant Messages	23	47	47
Download an application to your cell phone	17	26	24
Download or stream music or video	13	28	25

Source: Federal Communications Commission survey of 5,005 adult Americans, October-November 2009.

Not all of the activities listed in *Exhibit 14* constitute Internet use on a handheld. Texting is not an application that runs on the Internet while downloading maps and directions may require Global Positioning System

(GPS) capability. Not all downloading or streaming of music or video to a handheld requires Internet access.

In defining the scope of Internet use on a mobile handheld, the activities chosen for inclusion in the definition are: sending or receiving e-mail, accessing Web pages on the Internet and downloading an application to a cell phone. When those three items constitute accessing the Internet on a cell phone, 36 percent of all cell users have used the Internet on their devices. Since not everyone has a cell phone, this means 30 percent of all adults have gone online with a cell phone or smart phone.

As with the entire suite of handheld activities, minorities are more active in handheld Internet use than whites. This is how results break down across racial categories:

- 39 percent of Hispanics and of African-Americans have used the Internet on their mobile handheld device.
- 27 percent of whites have accessed the Internet on a mobile handheld.

A question to consider is if the differences are racial in origin or whether they can be explained by other factors, such as age or income, which may vary systematically across racial categories. White adults in the sample are, on the whole, older than the African-American respondents and much older than the Hispanics. The median age for white adults in the sample is 49; for African-Americans it is 46, and for Hispanics it is 33. As seen above, younger respondents are heavier users of mobile devices than older ones. While age, education and income are all indicators of the likelihood of using the Internet on a handheld device, the racial or ethnic classification of the respondent also emerges as a predictor–even when other factors are held constant. In other words, the differences that appear across races are not simply due to age or educational level.

Another issue is whether mobile Internet access is a substitute for home access. African-Americans and Hispanics report lower levels of home broadband adoption than whites yet are more likely to use the Internet on their handheld device *(See Exhibit 15)*.

Are lower home adoption rates attributable to mobile Internet use? The survey results indicate the answer is, for the most part, "no." There is a positive correlation between home Internet adoption and mobile Internet use–91 percent of mobile Internet users have Internet at home and 84 percent have broadband. Yet while African-Americans and Hispanics are more likely to be mobile Internet users, most users in these groups have broadband at home. More than three-quarters (78 percent) of African-American mobile Internet users have broadband at home, and more than two-thirds (68 percent) of Hispanic mobile Internet users have high-speed access at home.

Viewed from the perspective of nonbroadband adopters, few non-adopters access the Internet on their handheld devices, though African-Americans and Hispanics do this at a higher rate than average. Specifically, among those who do not go online at home with a broadband connection:

- 14 percent of all non-adopters have accessed the Internet on their mobile devices.
- 20 percent of African-American non-adopters have accessed the Internet on a mobile device.
- 25 percent of Hispanic non-adopters have accessed the Internet on their mobile devices.

Overall, there is not a high incidence of mobile Internet access replacing home Internet connection. Rather–since most mobile Internet users also have broadband–it is an additional pathway to access. This pattern is not quite as strong for African-Americans and Hispanics, but the general trend is, nonetheless, the same for those groups.

It is important not to confuse mobile Internet use with mobile broadband. To be sure, some of those using their mobile device for online access do so via relatively fast 3G and WiFi networks. However, it is difficult, using a survey, to determine what portion of mobile Internet users do this or how often.

Mobile Wireless Broadband

Another access pathway for people is mobile wireless broadband service on a laptop. The survey asked laptop users the following question to probe mobile broadband use:

> "Do you use a service with your laptop computer that is called wireless broadband, allowing you to access the Internet virtually anywhere? This is usually a service that you have to pay a monthly fee for, either by itself or as part of another communications bill. This is NOT what is called Wi-Fi."

Exhibit 15. By Race, Mobile and Home Internet Use

(% of mobile Internet users)		
	Internet at home	**Broadband at home**
White	94	89
African-American	92	78
Hispanic	78	68
Total	91	84
Number of cases	3,455	2,671

Source: Federal Communications Commission survey of 5,005 adult Americans, October-November 2009.

This access pathway appears at only modest levels in the general population. Among laptop users (52 percent of all adults), 28 percent said they used wireless broadband. As a percentage of the *entire* adult population, this comes to 15 percent who use wireless broadband.

African-Americans and Hispanics who have laptops were more likely than average to say they use such a service. Looking at the racial breakdown:

- Among African-Americans *with laptops,* 36 percent report using a mobile broadband service.
- Among Hispanics *with laptops,* 30 percent report using a mobile broadband service.
- Among whites *with laptops,* 26 percent report using a mobile broadband service.

The context for these findings is important. First, these figures apply to respondents with laptop computers–devices that Hispanics and African-Americans are less likely to have. (Some 54 percent of whites have a laptop, while 44 percent of African-Americans and 43 percent of Hispanics do.) As a share of the entire population of each of these groups, mobile broadband use unfolds as follows:

- Among all African-Americans, 16 percent say they use a mobile broadband service.
- Among all whites, 14 percent say they use a mobile broadband service.
- Among all Hispanics, 13 percent say they use a mobile broadband service.

Second, as noted, mobile broadband users are overwhelmingly home broadband users; 94 percent of all mobile broadband users have broadband at home, figures that are only modestly lower for African-Americans (91 percent) and Hispanics (84 percent). Although some may be using their mobile broadband connection as their principle home access means, those saying they use mobile broadband were as likely as the average to say they use DSL, cable modem service and other wireline means such as fiber. As with mobile Internet use, mobile broadband is mainly a supplementary broadband access pathway.

What People Pay for Cell Phone Service

The survey asked cell phone users what their monthly bill was, and the question included a prompt for respondents to include charges for texting or long-distance service. Only respondents who receive a stand-alone cell phone bill–80 percent of respondents–received this question. The average figure they gave was $92 per month.

By comparison, bill analysis from TNS Telecoms shows an average monthly cell phone bill of $99 per month. Many cell users have more than one line on their bill; for TNS, the average bill covered 1.9 lines. The FCC survey did not ask how many lines users had on their plan.

IV. Non-Adopters: Who They are and the Barriers They Face

Overview of Non-Adopters

Roughly one-third of Americans do not use broadband Internet at home. They fall into three categories of non-adoption:

- 22 percent of Americans do not use the Internet at all.
- 6 percent use the Internet but do not have access at home.
- 6 percent use dial-up Internet connections to go online from home.

Overall, 35 percent of Americans do not use broadband at home. (One percent of respondents were not broadband users, but could not be placed in any of three categories above). A small share–just 4 percent–say they cannot get service where they live. This is an imperfect measure of infrastructure availability, as it relies on the user accurately knowing about infrastructure in the neighborhood. This means that about one- third (31%) of Americans do not have broadband at home but could subscribe. However, the full cohort of non-adopters (35 percent) is included in our analysis and discussion.

Key demographic differences when contrasting all non- adopters with adopters are:

- **Gender:** 57 percent of non-adopters are women versus 49 percent of home broadband adopters.
- **People with disabilities**: 39 percent of non-adopters have a disability, compared with 15 percent of adopters.
- **College graduates**: Just 11 percent of non-adopters have college degrees versus 37 percent of broadband users.
- **Age**: 32 percent of non-adopters are age 65 or older versus 9 percent of adopters.
  - Nearly two-thirds (65%) of non-adopters who are senior citizens are women.
- **Income**: 43 percent of non-adopters live in households with annual incomes of $20,000 or less, compared with 17 percent of home broadband users.
- **Rural**: 24 percent of non-adopters live in rural areas versus 13 percent of broadband adopters.

Exhibit 16 presents detail on the demographic and socio-economic profiles for each of the three categories of non-adopters. Non-Internet users, the largest group, are older, lower-income and the least educated of the three. Online users who do not have home access tend to be younger women, usually with no more than a high school degree and low incomes. Dial-up users, though not as well off economically as broadband users, are higher on the socioeconomic scale than the other groups of non-adopters.

Past and Proxy Broadband Use for Non-Adopters

Although those classified as non-adopters are not broadband users at home today, this does not mean that all of them are completely cut off from modern ICT goods or services. Many have a working computer at home (discussed below) and 49 percent uses computers occasionally, either at home, work or school.

Exhibit 16. The Demographics of Non-adopters

	Broad-band at home	Dial-up at home	Internet users, not at home	Non-Internet users
Gender				
Male	51%	49%	41%	45%
Female	49	51	59	55
Families				
Parents with minor children at home	36	24	27	19
Disabled				
Those who report they have a disability	15	19	28	47
Age				
18-29	25%	16%	39%	11%
30-49	39	27	34	20
50-64	26	33	20	28
65+	9	24	8	41
Median Age	43	53	38	60
Race/ethnicity				
White (not Hispanic)	74%	75%	64%	63%
Black (not Hispanic)	10	10	15	13
Hispanic (English or Spanish speaking)	10	10	15	20
Educational attainment				
Less than high school	5%	12%	8%	39%
High school graduate	29	37	50	45
Some college	29	28	23	11
College +	37	23	19	5
Household income				
Under $20K	10%	16%	30%	32%
$20K-$30K	7	13	16	17
$30K-$40K	9	11	18	7
$40K-$50K	9	15	7	8
$50K-$75K	16	14	11	6
$75K-$100K	15	8	6	1

	Broad-band at home	Dial-up at home	Internet users, not at home	Non-Internet users
Over $100K	20	5	3	2
Don't know/refused	15	19	10	27
Community type				
Urban	31	24	22	31
Suburban	51	43	40	45
Rural	14	29	21	22
Number of cases	2,671	459	392	1,450

Source: Federal Communications Commission survey, October-November 2009. Total number of cases in this table is 4,972. This does not reflect the entire sample size of 5,005 because some responses did not permit classification into any of the categories.

Additionally, some have had experience with broadband. Among dial-up users or "not-at-home" users, 46 percent have used broadband at some point– maybe at work, perhaps a friend's house or possibly somewhere else. Among the wider set of all non-Internet users, 17 percent in this group were once Internet users at home. Of these former users, 49 percent say broadband was the type of home connection they had.

Combining these two sets of nonusers' past broadband experience yields reveals that one-quarter of non-adopters have had some experience with broadband–meaning they once had broadband access at home or have used it at work, a friend's house *or* some other place. Put differently, 8 percent of the general population have had some exposure to broadband but do not presently have it at home.

As to prospects for moving nonusers to broadband, 20 percent of nonusers say they would be interested in getting online access. Among current "not-at-home" Internet users, 46 percent would be interested in getting home access, while 41 percent of dial-up users say they would be interested in getting broadband. Only 17 percent of dial-up users say they are very likely to get broadband in the next year. Similarly, among *all* interested nonusers, broadband access is not a sure bet in the near future. Out of all non-adopters, 26 percent say it is very likely they will get broadband in the next 12 months, with another 28 percent saying this is somewhat likely.

There is also the phenomenon of proxy access, that is, nonusers who cohabit with another Internet user. Among respondents who are non-Internet users or who do not use the Internet at home, 22 percent live with someone who uses the Internet at home. These nonusers often ask their online housemates to carry out tasks online for them. One in six (16 percent) say they ask their online housemates to do something online for them once a week. Another 20 percent say they do this once a month, while an additional 24 percent say they do a few times a year.

A portion of the proxy access that nonusers' housemates carry out is broadband. Nonusers who live with someone who goes online from home were asked about the household's access. Some 21 percent say it is a dial-up connection, 47 percent state it is some other way and 32 percent do not know. Assuming that the response "some other way" reflects a home high-speed connection, these questions suggest that 2 percent of the general population fall into the category of nonusers who live in a broadband-connected household.

The Role of Disabilities in Non-Adoption

The largest group of nonbroadband adopters–non-Internet users–has a high likelihood of having some sort of disability. Overall, 39 percent of non-adopters have some sort of disability. The survey posed six questions, which, if any one yielded a "yes" answer, classified the respondent as a disabled American *(See Exhibit 17)*.

Collectively, 24 percent of respondents had an affirmative response to one of those questions and, for purposes of this report, are classified as people with disabilities.[18] Within this group, 42 percent have broadband at home–much lower than the 65 percent average. Some 56 percent of those with disabilities are Internet users, below the 78 percent average. As the exhibit presenting the demographic and socioeconomic profiles of non-adopters shows, nearly half (47 percent) of non-Internet users report having some sort of disability.

It is important to note that people with disabilities and non-adopters overlap quite a bit and in fact share many characteristics with non-adopters: They are generally older and have lower incomes. For that reason, it is not a surprise that non- adopters include a disproportionately high share of people with disabilities. Some of the difference in adoption rates is due to individuals' disabilities and some is due lower incomes, advanced age or other factors associated with low adoption.

Exhibit 17. Questions Used to Determine Disability (by Percentage)

Do you have serious difficulty walking or climbing stairs?	12
Because of a physical, mental or emotional condition, do you have serious difficulty concentrating, remembering or making decisions?	8
Are you deaf or do you have serious difficulty hearing?	8
Because of a physical, mental or emotional condition, do you have difficulty doing errands alone such as visiting a doctor's office or shopping?	6
Are you blind or do you have serious difficulty seeing even when wearing glasses?	5
Do you have difficulty dressing or bathing?	2

Source: Federal Communications Commission survey of 5,005 adult Americans, October-November 2009.

Why Non-Adopters Do Not Have Broadband

The survey sought to determine with as much specificity as possible why people without broadband choose not to have the service at home. Respondents were read a list of possible reasons for not having broadband and permitted to list as many as they chose as barriers. They then received a follow-up question asking the *most important* reason they did not have broadband.

The question wording and choices offered varied slightly across different non-adopter groups. Dial-up users, for example, were offered the choice of "happy with current service," which is not a relevant choice for those who do not use the Internet at all. A number of the exhibits show how people responded to the question that allowed them to list multiple reasons for non-adoption, then how they responded to the question asking about the most important reason.

Non-Internet users–22 percent of the adult population

Those who do not use the Internet offer a range of reasons for not being online. Nearly half cite the monthly cost, and similar numbers point to comfort with computers or worries about online content–that is, digital literacy looms as a barrier for them. The affordability of the computer is also a prominent reason for nonusers *(See Exhibit 18).*

Exhibit 18. Reasons Nonusers Do Not Use the Internet (by Percentage)

Monthly cost is too expensive	47
I am not comfortable using a computer	46
I am worried about all the bad things that can happen if I use the Internet	45
The activation and installation fee to get service is too much	42
I cannot afford a computer	40
There is nothing on the Internet I want to see or use	35
The Internet is just a waste of time	33
I can access the Internet all I need to at work	14
It's not available where I live	13

Source: Federal Communications Commission survey, October-November 2009; n=1,450 for nonusers.

The survey permitted respondents to choose more than one reason from the list of barriers to adoption. Fifty percent of nonusers picked at least three reasons, and one-quarter (24 percent) cited at least five of the reasons listed.

When asked the most important reason for not using the Internet, no single reason stands out for nonusers. About 16 percent cite lack of comfort in using a computer as the main barrier, but cost (either monthly fee, cost of computer or activation fee) is the main reason cited by 25 percent of dial-up users. Some 12 percent are worried about risks online, and 14 percent cite factors that suggest they are unclear about the Internet's relevance to them (with 7 percent saying the Internet is a waste of time and 7 percent saying there is nothing they want to see online). *(See Exhibit 19)*

Those who cited monthly cost as a barrier (47% in this category of non-users) received follow-up questions on what they would be willing to pay for service. The responses sorted into three categories:

- 40 percent of nonusers provided an estimate and, among this group, the average monthly figure they said they would pay for Internet access at home was $26; the median figure was $20.
- 31 percent said either they don't know or refused to answer.
- 29 percent said they were not willing to pay anything for access.

As with nonusers, it is common for dial-up users to cite more than one reason that keeps them from using the Internet. Fifty- six percent cite at least three reasons and 18 percent cite five or more *(See Exhibit 20).*

Exhibit 19. Main Reasons Nonusers Do Not Use the Internet (by Percentage)

I am not comfortable using a computer	16
I cannot afford a computer	14
I am worried about all the bad things that can happen if I use the Internet	12
Monthly cost is too expensive	11
The Internet is just a waste of time	7
There is nothing on the Internet I want to see or use	7
The activation and installation fee to get service is too much	5
It's not available where I live	2
I can access the Internet all I need to at work	1
None of above reasons	9
Combination of reasons	5
Other reason	5
Don't know/refused	3

Source: Federal Communications Commission survey, October-November 2009; n=1,450 for nonusers.

About half of dial-up users are satisfied with their service or are not heavy Internet users. Half also cite the monthly cost, with a significant share saying the installation fee is a barrier. Nearly one-third say broadband is not available where they live. However, it is important to put this figure in proper context. First, dial-up users make up 6 percent of the adult population; if 30 percent truly cannot get broadband where they live, this means just 2 percent of adults (from this set of respondents who received the question) lack broadband availability. Second, since this is self-reported data on infrastructure availability, it is possible that respondents are wrong as to whether broadband is available where they live. Nonetheless, most (74 percent) of dial-up users who say availability is a barrier report that they would get broadband if it were available in their area.

When probed further on the *most important* barrier, issues pertaining to the relevance of the service (that is, not using the Internet very much, being content with current dial-up service and not needing additional speed) together loom largest (38percent) for dial-up users. Cost issues–monthly fee, activation fee, unwillingness to enter into a long-term contract–are next, cited as most important by 28 percent of dial-up users.

Dial-up users who cited cost as a barrier then received a follow-up question asking how much they would be willing to pay for a broadband connection at home *(See Exhibit 21).* As with nonusers, responses fell into three categories. Among dial-up users citing monthly cost as a reason they do not have broadband:

- 62 percent provided an estimate. The average amount this group was willing to pay for broadband was $28 per month and the median figure was $25.
- 25 percent said they did not know or refused to answer the question.
- 13 percent said they would not be willing to pay anything for broadband.

Exhibit 20. Reasons Dial-Up Users, Who Make Up 6% of the Adult Population, Have Not Switched to Broadband (by Percentage)

The monthly cost is too expensive	50
I'm happy with my current service	49
I do not use the Internet that much	44
I do not want to enter into a long-term service contract	44
The activation or installation fee to get service is too much	44
I do not need the additional speed it would offer	31
It's not available where I live	29

Source: Federal Communications Commission survey, October-November 2009; n=567 for home dial-up users or those who did not know connection type.

Exhibit 21. Reasons Dial-Up Users Have Not Switched to Broadband (by Percentage)

The monthly cost is too expensive	19
I do not use the Internet that much	18
It's not available where I live	17
I'm happy with my current service	16
I do not want to enter into a long-term service contract	4
I do not need the additional speed it would offer	4
The activation or installation fee to get service is too much	5
None of these reasons	7
Other/Don't know	2

Source: Federal Communications Commission survey, October-November 2009; n=567 for home dial-up users or those who did not know connection type.

Not-at-home users–6 percent of the adult population

Although the vast majority of Internet users have the means to go online from home, 6 percent do not. They use the Internet from elsewhere–perhaps work, perhaps the library–but not where they live.

For "not-at-home" users, worries about comfort with a computer or thinking there is nothing worthwhile online are not big barriers. Rather, affordability (e.g., monthly fee, activation charge or computer cost) plays a larger role. While those reasons are front and center, "not-at-home" users cite multiple reasons why they are not online. Two-thirds (66 percent) list at least three reasons and 11 percent cite five or more *(See Exhibit 22).*

Affordability clearly comes to the forefront when "not-at-home" users are asked about the most important barrier they face. Some 41 percent cite the monthly cost or the outlay for a computer as the reasons they do not use the Internet from home.

As with prior groups, "not-at-home" users who said monthly cost was a barrier received a follow-up question on what they would be willing to pay for Internet access at home. Among "not-at-home" users for whom cost is a barrier *(See Exhibit 23)*:

- 74 percent gave a dollar estimate; the average figure cited as this group's willingness to pay for broadband was $21 per month and the median figure was $20.
- 16 percent said they did not know or refused to answer the question.
- 10 percent said they would not be willing to pay anything for broadband.

What Non-Adopters are Willing to Pay

Across the three categories, 51 percent of all non-adopters said monthly cost was a barrier, although fewer said it was the *primary* reason they lack service. As is evident from the preceding discussion, not all of those who said cost was a barrier were able to respond to follow-up questions on how much they would be willing to pay for service. Across all three categories, just more than half (52 percent) of those who said monthly cost was a barrier could provide an estimate to the nearest dollar of what they would pay for service. As a share of all non-adopters, 26 percent provided an estimate of willingness to pay for service. The willingness-to-pay figure for all non-adopters was $25 per month.

The $25 average does not mean that all non-adopters asked the question would purchase service if they faced a price at that level. Some respondents cited a figure above $25, many cited a figure below. In fact, among those who offered an answer, 65 percent cited a figure of $20 per month or more for WTP. The vast majority (91 percent) of those who gave an answer cited a figure of $10 per month or more.

Taking these respondents at their word–that they really would get broadband at the price they state–and representing the figures above for $20 per month and $10 per month as a share of *all* adults:

- A $20 offer would yield an increase in overall broadband adoption of 6 percentage points.
- A $10 offer would yield an increase of 8 percentage points.

These small impacts reflect the fact that the survey tested the hypothesis that there are multiple reasons for non-adoption, with cost being one among them. The survey found that people had multiple barriers with, as noted previously, 51 percent citing cost as one of them, followed up by a question on willingness-to-pay. Just more than half cited a figure, while 28 percent answered "don't know" and the balance said they would pay nothing for broadband.

Exhibit 22. Reasons "Not-at-Home" Users Have Not Switched to Broadband (by Percentage)

Monthly cost is too expensive	57
The activation and installation fee to get service is too much	51
I can access the Internet all I need to at work	34
I cannot afford a computer	32
I am worried about all the bad things that can happen if I use the Internet	31
It's not available where I live	15
There is nothing on the Internet I want to see or use	14
The Internet is just a waste of time	14
I am not comfortable using a computer	13

Source: Federal Communications Commission survey, October-November 2009; n=392 for not-at-home users.

Exhibit 23. Reasons "Not-atHome" Users Have Not Switched to Broadband (by Percentage)

Monthly cost is too expensive	27
I cannot afford a computer	15
I can access the Internet all I need to at work	11
I am worried about all the bad things that can happen if I use the Internet	9
The activation and installation fee to get service is too much	6
It's not available where live	5
I am not comfortable using a computer	5
The Internet is just a waste of time	4
There is nothing on the Internet want to see or use	2
None of these reasons	7

Source: Federal Communications Commission survey, October-November 2009; n=392 for not-at-home users.

Finally, it is important to underscore the uncertainties inherent in measures of willingness-to-pay. Respondents are asked to make judgments on the value of a service that, many at least, have used sparingly (if at all); they therefore lack information on service attributes, which is key to placing a dollar value. For that reason, measures of willingness to pay, and the scenario on how much additional adoption would ensue from specific price offerings, should be taken as illustrative, not predictive.

Summary on Most Important Reason for Non-Adoption

The three baskets of non-adopters were asked similar kinds of questions about why they do not have broadband or any Internet service at home. Pulling the reasons together across all three groups of non-adopters yields the following consolidated reasons for non-adoption among ***all*** those who do not have broadband at home.

Cost: 36 percent of non-adopters cite a cost-related reason as their main barrier to adoption, with 15 percent pointing to the monthly cost of service, another 10 percent citing affordability of a computer and 9 percent saying they do not want a long-term service contract or find the installation fee too high. The remaining 2 percent cited a combination of these reasons.

Digital Literacy: 22 percent of non-adopters give a digital literacy-related topic as their main barrier. These are non- adopters who said that "they are not comfortable using a computer" or they "are worried about all the bad things that can happen on the Internet." Digital literacy breaks down as follows: 10 percent of non-adopters cite "worries about bad things" and 12 percent said they are "not comfortable with computers." Although worries about online hazards (e.g., exposure to inappropriate content or the possibility of identity theft) exist for many current broadband users, some non-adopters are concerned enough to avoid the Internet altogether. This may have to do with the age of those citing digital literacy as a barrier. Their median age is 62.

Relevance: 19 percent of non-adopters say they do not think digital content delivered using broadband is sufficiently compelling to justify getting it. Many view broadband as an avenue to irrelevant content, and others seem content with the offline alternatives currently available to them. These respondents say, for instance, that the Internet is a "waste of time," they do not believe there is anything worth viewing online or, for dial-up users, they cite contentment with their current service. Each of those reasons indicates that these non-adopters are on the other side of a perceptual chasm with respect to broadband. Unlike broadband users, they are not attuned to online content's potential to provide information or opportunities for learning.

Other: 11 percent of non-adopters cited a variety of reasons that did not group into an identifiable category or offered no response.

Combination: 4 percent of non-adopters cited as their main reason a combination of reasons listed, whether that was price, worries about "bad" things online or other items.

Lack of availability: 5 percent of non-adopters said that the main reason they do not have the Internet or broadband is that it is not available where they live. This response is highest among dial-up users, 21 percent of whom say that broadband is not available where they live.

Use high-speed at work: Finally, 3 percent of non-adopters said the main reason they do not have broadband at home is because they use broadband as much as they want at the workplace.

For the "lack of availability" reason, it is also important to distinguish respondents citing lack of availability as the most important reason for not having broadband from those who cited it as one of several reasons. Because people could list multiple reasons for not adopting (prior to being asked the most important reason), some who cited lack of infrastructure availability could (and did) not cite that as their most important reason for non-adopting. The upshot is that 12 percent of non-adopters cited lack of availability as a reason for not having broadband, while 5 percent of non-adopters cited it as the most important reason. The 12 percent of non-adopters citing lack of available infrastructure translates into 4 percent of all Americans who say they cannot get broadband because it is not available where they live. Finally, it is important to underscore that this is a self-reported assessment about the availability of broadband infrastructure, not an actual measure of whether broadband is in the respondent's neighborhood.

When Respondents Can Cite Multiple Reasons for Not Adopting

It is also worth summarizing, for all non-adopters, what share cited any of the reasons listed as possible barriers. The survey permitted non-adopters to list several reasons for not adopting before asking them to state the most important reason. Because respondents could list more than one reason–in fact most cited multiple reasons–the figures below sum to more than 100 percent.

Cost: 66 percent of non-adopters cited any of the reasons that constitute the "cost" category, that is, two-thirds of non- adopters listed monthly access fee, cost of computer, installation fee, or reluctance in entering into a long term contract, as a reason they do not use the internet or broadband.

Relevance: 52 percent of non-adopters said that they found the internet to be a waste of time, they did not think there is anything worth seeing online, or they were content with their current service.

Digital literacy: 47 percent of non-adopters listed their lack of comfort with computers or worries "about all the bad things that can happen on the internet" as reasons for not having broadband.

Lack of availability: 12 percent of non-adopters said that their inability to get service where they live is why they do not have broadband.

To underscore how barriers to access interact in users' minds, the following shows the frequency with which those who cited a particular barrier also pointed to others.

Among the 66 percent of non-adopters who cited cost as a barrier:

- 50 percent also listed a digital literacy-related barrier.
- 48 percent also listed concerns about the relevance of the internet as a barrier.

Among the 52 percent of non-adopters who listed lack of relevance as a barrier:

- 45 percent also listed a digital literacy-related barrier.
- 28 percent also listed on the cost-related reasons.

Among the 47 percent of non-adopters who listed digital literacy as a barrier:

- 70 percent also listed cost as a barrier.
- 57 percent also listed lack of relevance as a barrier.

An important consequence of this analysis is that non- adopters typically face multiple barriers to adoption. Even though cost leads the way, most of the time even those who worry about cost need help in other areas. Likewise, those who say lack of relevance is a barrier also have issues with digital literacy and, to a less extent, cost.

Segmenting the Population of Non-Adopters

Each of the three baskets of non-adopters reveals tendencies as to why people do not have broadband. Financial matters are critical for not-at-home Internet users, while dial-up users seem distant from the Internet, apparently not seeing a lot online that might lure them to faster service. Nonusers–the largest group of non-adopters, constituting two-thirds of non-adopters–are harder to pin down. Many cite reasons suggesting they do not see the relevance of the Internet, yet some point to costs and others worry about risks in being online.

Even if the survey responses yielded crisp reasons for non-adoption across categories of nonbroadband users, it would be worthwhile to know more than just a person's stated barrier to adopting. In particular, it would help to know something about the person's ability to clear the barrier.

One technique to do this is to segment the population of non-adopters into several groups based on measures of a respondent's relationship to the Internet. The following analysis does this and it is built on two dimensions:

- **Proximity to ICTs**: All survey respondents were asked whether they have computers, cell phone or premium TV at home. Nonbroadband adopters also were asked if they had ever used a computer.
- **Attitudes about the Internet and computers**: Respondents received questions about their overall comfort with computers and attitudes about the Internet.

The survey details the specific attitudinal questions and responses for broadband adopters and non-adopters. It is clear that people without broadband at home use ICTs. Among non-adopters:

- 80 percent have premium TV, either cable or satellite television.
- 70 percent have a cell phone.
  - On average, these cell phone users spend $73 per month on service.
- 49 percent say they use computers, at least occasionally.
- 42 percent have at least one working computer at home.
- Among nonbroadband-adopting computer users, 34 percent say they are *very* comfortable using them.

On the second issue, non-adopters have a mix of attitudes about the Internet–some positive, others less so. *Exhibit 24* compares "strongly agree" responses for broadband adopters and non-adopters.

The attitudinal and proximity questions serve as input into statistical analysis that classifies respondents based on their responses. That analysis includes all non-adopters–dial-up, not-at-home users and nonusers. The analysis yielded four groups of non-adopters, each with distinct characteristics described below. The analysis also identifies the key adoption barriers in each group, although answers to those questions were not inputs into the statistical analysis that created the groups.[19]

Group 1. Digitally Distant: 10 percent of the general population

This group does not see the point of being online and does not have the computer skills to dive in even if its members were so inclined. Few (15 percent) in this group see the Internet as a tool for learning and most see it as a dangerous place, either for children (53 percent) or personal information (56 percent). Very few (11 percent) have a computer. Many (25 percent) cite digital literacy as a barrier to adoption, and nearly half are retired from the work force. Just 8 percent of this group have any experience with broadband (i.e., they used at a place other than home or used to have home high-speed service). The Digitally Distant group has a *low probability of adoption* in the near future.

The important demographic characteristics of this group have to do with age. With a median age of 63, it is the oldest of any of the groups; 47 percent are age 65 or older. The Digitally Distant also has the highest rate of retirees across the four groups, at 45 percent. They are also not well off economically, with 50 percent reporting annual household incomes below $30,000.

Group 2. Digital Hopefuls: 8 percent

The Digital Hopefuls like the idea of being online, but lack the resources to connect using broadband. Members of this group have positive attitudes about how the Internet might be a tool for learning and personal productivity; 86 percent strongly believe it is a useful tool for learning and 68 percent believe people can be more productive using the Internet. But 22 percent cannot afford a computer (and only 3 percent have one at home), 15 percent cannot afford the monthly access fee and only 9 percent have past experience using broadband. A significant share (28 percent) cites digital literacy as a barrier, the highest of any group. Given help–likely a lot of it–in clearing cost hurdles to access, members of this group have a *positive probability* of adopting broadband.

Consistent with their lack of resources to connect, the Digital Hopefuls are the least well off economically of any group. Some 44 percent live in households with annual incomes of $20,000 or less. Nearly half (49 percent) have not graduated from high school. This group, with a median age is 57, has the highest share of Hispanics (26 percent). One in five are African-Americans and 50 percent are white.

Exhibit 24. How Strongly Do You Agree or Disagree With the Following Statements (% of Non-adopters Who "Strongly Agree")

There is too much pornography and offensive material on the Internet	65
It is too easy for my personal information to be stolen online	57
The Internet is a valuable source for information and learning	59
It is important for children to learn how to use the Internet	54
The Internet is too dangerous for children	46
People can be more productive using the Internet	37

Source: Federal Communications Commission survey, October-November 2009; n=2,334 for non-adopters.

Group 3. Digitally Uncomfortable: 7 percent

The Digitally Uncomfortable has the equipment for access (nearly all have a computer), but the group has tepid attitudes about the usefulness of the Internet and low comfort levels with computers. Just 24 percent say they are very comfortable using them, and just more than one-third (37 percent) see the internet as a productivity tool. Although some in the Digitally Uncomfortable group could use a break on the monthly access fee (16 percent), more need help learning about relevant content (18 percent) and improving their digital literacy (17 percent). Twenty-one percent cite a lack of available infrastructure as a reason they do not have broadband.

As to attitudes, 53 percent worries about security of personal information online, although a majority (68 percent) sees the Internet as a useful tool for learning and information gathering. The path to access and robust use might be long for this group. Because they have already cleared the computer ownership hurdle and because one-third (35 percent) have past experience with broadband, the Digitally Uncomfortable, with the right support, *stand a reasonable chance of adopting broadband.*

From a socioeconomic perspective, members of the Digitally Uncomfortable group are very much the average non-adopter. The median age is 55 (versus 54 for all non-adopters), 63 percent are white (the same as the average), and 27 percent have either graduated from college or at least attended college (also the average).

Group 4. Near Converts: 10 percent

This group has many of the qualities of broadband adopters– a high rate of computer ownership (76 percent use them occasionally and 68 percent own one), comfort with computers (56 percent describe themselves as very comfortable with the device) and positive attitudes about the Internet's usefulness (three-quarters say it is a valuable learning tool). And 42 percent have used broadband in the past. However, nearly 41 percent cite financial issues as the main barrier to home broadband access, principally the monthly access fee. Concerns about the Internet's relevance to their lives also play a role. At 10 percent of the general population, this group has a *high probability of adopting broadband*–if they clear financial constraints.

The Near Converts are much younger than other non-adopters (their median age is 45) and they have much higher levels of educational attainment. Half have either graduated from college or attended college. Their incomes are in the middle ranges (30 percent have annual household incomes between $40,000 and $75,000) and most are employed full or part time.

Exhibit 25 summarizes the main barriers that each segment of non-adopters faces. A clear implication of *Exhibit 25* is this: The greater the conversion probability, the larger role the monthly Internet bill plays as a barrier. The Near Converts have the highest probability of becoming broadband adopters; the monthly access cost is the main hurdle. The most difficult-to-convert group is the Digitally Distant; they have no single reason standing out as a barrier. In between the two extremes, cost plays a role in slightly different ways (computer affordability for the Digitally Adrift, monthly bill for the Digitally Disposed), but other factors are important, too.

At the same time, although cost leads as a primary barrier for non-adopters, nearly two-thirds cite a reason other than cost. As noted earlier, non-adopters generally mention more than one obstacle for adoption–meaning interventions to lure people to broadband should be comprehensive in nature. Cost relief will work effectively for many non-adopters but only when accompanied by training programs to bolster their digital skills and information about content that is relevant to their lives. Similarly, addressing digital literacy in isolation is not likely to be effective unless cost is addressed and efforts are made to inform people about online content relevant to them.

V. FOCUS ON KEY POPULATION GROUPS

Although this paper has referred to differences among population subgroups, this section has key figures for several groups of interest, as well as discussion of notable variations in behavior and attitudes for these groups. This focus on specific groups covers data on adoption and non-adoption.

Exhibit 25. Summary of Most Important Barriers by Segment

Segment	% as share of all Americans today	Adoption barriers faced
1. Near Converts	10	Cost (mostly monthly bill)
2. Digital Hopefuls	7	Cost (mostly monthly bill), relevance
3. Digitally Uncomfortable	8	Cost (mostly PC), digital literacy, accessibility
4. Digitally Distant	10	Cost, digital literacy, relevance, accessibility
Total Non-adopters	**35**	

Families (Respondent is the Parent of Child Under 18 Who Lives in Household)

With the integration of information technology into so many dimensions of family life–whether is communicating with one another or schools–it is no surprise that those with children under age 18 in the house generally have a lot of information technology. Nine out of 10 (87 percent) have Internet access, and 74 percent have broadband at home. The survey interviewed parents, of whom 92 percent have a cell phone, and many (39 percent) use it sometimes to access the Internet. Among kids under 18 living at home, 45 percent have cell phones. For children between the ages of 12 and 17, some 75 percent have cell phones.

Although families with minor children at home have higher broadband adoption rates, there is still a gap between broadband adoption and computer ownership. Among such families, 87 percent have a working computer at home–a 13-percentage-point gap compared with 74 percent of those without children who have a working computer at home.

The survey asked parents or guardians the ages of their children. When it came broadband adoption, there was little difference between parents with children from ages 5 to 12, and those with children between 13 and 17; the level was 72 percent for families with younger children and 75 percent for families with older ones. To the extent large adoption differences unfold by age, younger parents are less likely to have broadband at home than older ones. For parents with offspring at home in the 18-to-29 age range, 65 percent have broadband at home, compared with 77 percent of parents older than that.

As to online activities, parents tend to engage in activities that focus on convenience and community. Fully 86 percent have used their broadband connection to keep up with community news, 85 percent for online purchases and 57 percent to get advice or information from a government agency about a health or safety issue; all of these exceed the average. When asked what activities are *very* important to them, 44 percent cite keeping up with news about the community; that is above the average.

Non-adopters

Cost–and the monthly fee in particular–comes front and center as the barrier to adoption for non-adopting families. Half (48 percent) say that cost is the primary reason they are not online, with 24 percent specifically citing the monthly. Issues such as digital literacy (16 percent) or lack of relevance (13 percent) figure much less. Note that the "other" category in *Exhibit 26,* and ones that follow, refers to all other reasons cited by respondents, which may include items such as lack of availability and ability to use the Internet from work.

Low-Income Americans

There is a clear relationship between broadband adoption and income. Some 40 percent of those with household incomes of $20,000 per year or less have broadband at home, while 91 percent of those with household incomes above $75,000 per year have it *(See Exhibit 27).*

Exhibit 26. Barriers to Adoption: Families with Minor Children at Home (by Percentage)

Main reason cited for not having the Internet or broadband	Families with minor children at home	Other non-adopters
Cost	46	31
Digital literacy	16	24
Relevance	13	21
Other	25	25

Source: Federal Communications Commission survey, October-November 2009.

Exhibit 27. By Income, Reasons for Broadband Use (by Percentage)

	Low-income broadband users (annual household income of $20K or less)	Upper-income broadband users (annual household income of $75K or more)
Activities in which low-income users lead		
Get information about a job or apply for one	77	60
Play games	70	44
Stream music to your computer	64	55
Take a class for credit	31	25
Play complicated role-playing games	21	9
Activities in which upper-income users lead		
Buy a product	74	93
Go to local, state, or government Web site	68	86
Get international news	64	85
Banking	51	80
Review a product or service	49	62
Contact a government agency about health or safety issue	42	62

Source: Federal Communications Commission survey, October-November 2009.

Not only does broadband adoption vary by income, individuals with broadband at home exhibit some differences in behavior and attitude toward broadband depending on their income. In general terms, lower-income broadband users are more likely to use their home high-speed connections to address important life issues, such as job searches or education, and for entertainment. Higher-income broadband users, however, are more likely than low-income ones to shop online, contact government and bank *(See Exhibit 27)*.

When asked whether particular online activities were very important to them, low-income and upper-income broadband users share the idea that communicating with family and friends is very important. By a modest margin (74 percent to 68 percent), low-income broadband users are more likely to say that than upper-income ones. Low-income users are also more likely to say that entertainment activities are very important to them. One in five (21 percent) low-income broadband users say that watching TV, movies or videos online is very important to them, compared with 8 percent of upper-income users who say this. However, upper-income broadband users are nearly twice as likely to say that shopping online is very important to them; 32 percent say this versus 17 percent of low-income broadband users.

Non-adopters

Unsurprisingly, those with the lowest incomes are most likely to cite cost as the main barrier to having broadband at home *(See Exhibit 28).*

For the other barriers–digital literacy and relevance–the stories are mixed. For digital literacy, aside from low-income people citing it less often, around one-fourth of non-adopters point to it as a primary barrier. The Internet's perceived lack of relevance is more of an issue in the middle-income ranges. For higher-income non-adopters (of which there are few), hard-to categorize "other" reasons dominate.

African-Americans

For African-Americans, online access is, relative to the average, somewhat less oriented to home high-speed wireline access and more focused on mobile Internet. Nearly three in five (59 percent) African-Americans have broadband at home, but 39 percent have used the Internet on their mobile handheld device. For African-Americans, home broadband adoption trails the national average by six percentage points; for mobile Internet use, African-Americans outpace the national average by nine percentage points.

Exhibit 28. Reasons for Not Adopting Broadband (by Percentage)

	Cost	Digital literacy	Relevance	Other
Under $20K	47	17	15	21
$20K–$30K	37	23	18	23
$30K–$40K	38	21	14	27
$40K–$50K	29	24	23	25
$50K–$75K	25	23	26	23
Over $75K	25	20	12	43

Source: Federal Communications Commission survey, October-November 2009.

To an extent greater than average, African-Americans without broadband at home use the mobile Internet as a substitute, but it is not a very widespread practice. Among African- Americans who have gone online with their mobile device, 78 percent have broadband at home; for all mobile Internet users, the home broadband adoption rate is 89 percent. For African-Americans *without* broadband at home, 20 percent have used the Internet on their handheld devices; that figure is just 9 percent for all non-adopters.

Another distinctive pattern among African-Americans is that broadband adoption gaps are particularly acute for older African-Americans. In fact, broadband adoption for African-Americans is 76 percent in the 18–29 age group, essentially the same others. Thereafter, however, African-Americans are much less likely than average to adopt broadband at home when the focus is on age and the gap is most pronounced among senior citizens *(See Exhibit 29).*

Because of the high share of young people in the ranks of African-American broadband adopters, the online activities of African-American broadband users takes on a youthful cast. Some 83 percent of African-Americans with broadband have used the connection to search for or apply for a job, far above the 57 percent rate for other broadband users. African-American broadband users are also more likely to take a class for credit online or participate in a blog *(See Exhibit 30).*

As to what they see as very important, African-American broadband users are very likely to say communicating with family and friends, although at a somewhat lower rate than average. African-Americans are, however, more likely to say that the Internet is very important to them for keeping up with community news and entertainment *(See Exhibit 31).*

Exhibit 29. Broadband Adoption by Age: African-Americans (by Percentage)

	African-Americans	All others in sample
All	59	66
Ages 18–29	76	75
Ages 30–49	67	75
Ages 50–64	49	66
Ages 65+	21	36

Source: Federal Communications Commission survey, October-November 2009.

Exhibit 30. Online Activities: African-Americans (by Percentage)

	African-Americans	All other broadband adopters
Get information about or apply for a job	83	57
Update a blog (either own or group blog)	37	24
Take a class toward a degree	37	22

Source: Federal Communications Commission survey, October-November 2009.

Exhibit 31. Online Attitudes: African-Americans

% who say this is *very important* to them	African-Americans	All other broadband adopters
Communicating with family & friends	61	67
Keeping up with news in the community	51	38
Watching TV shows, movies or other video	18	9

Source: Federal Communications Commission survey, October-November 2009.

Non-adopters

For African-Americans without broadband access at home, cost is cited as the main reason–and it weighs in at a rate somewhat higher than average. Forty-two percent of African-Americans cite cost as the primary reason they do not have broadband at home; this compares to an average 36 percent among non- adopters *(See Exhibit 32).*

Dissecting the cost factor shows that 15 percent of African-Americans point to the affordability of a personal computer (above the 10 percent average), and 15 percent cite the level of the monthly bill (right at the average).

Hispanics

The gap separating Hispanics' home broadband use from the average is wider than the comparable figure for African-Americans; 49 percent of Hispanics have broadband at home. There is also a gap in adoption within the groups, as Spanish- speaking Hispanics are far less likely to have broadband at home than those whose facility with English is such that they opted to take the survey in English. For Hispanic respondents who chose to take the survey in

Spanish, just 20 percent have broadband at home. Hispanics who took the survey in English are right at the 65 percent average for home high-speed adoption *(See Exhibit 33)*.

The adoption gap for Hispanics is most acute among young adults. Some 57 percent of Hispanics between the ages of 18 and 29 have broadband, more than 20 points below the average. Just more than half (53 percent) of those between the ages of 30 and 49 have broadband, 25 points behind the average.

Hispanics are not laggards when it comes to the cell phone. The vast majority (85 percent) has one and, along with African- Americans, members of this group are active users of the mobile Internet; 39 percent have taken advantage of online access with their handheld. To a somewhat greater extent than African-Americans, mobile access takes the place of broadband at home. One-quarter of Hispanics who *do not* have broadband at home access the Internet using their mobile device. Among Hispanics who have used the Internet via handheld devices, 68 percent have broadband at home, another indication that the mobile Internet fills the wireline void for some Hispanics.

Exhibit 32. Barriers to Adoption: African-Americans (by Percentage)

Main reason cited for not having the Internet or broadband	African-Americans	Other non-adopters
Cost	42	35
Digital literacy	20	22
Relevance	13	20
Other	25	24

Source: Federal Communications Commission survey, October-November 2009.

Exhibit 33. Broadband Adoption by Age: Hispanics (by Percentage)

	Hispanics	All others in sample
All	49	67
Ages 18–29	57	81
Ages 30–49	53	78
Ages 50–64	32	67
Ages 65+	23	36

Source: Federal Communications Commission survey, October-November 2009.

Exhibit 34. Barriers to Adoption: Hispanics (by Percentage)

Main reason cited for not having the Internet or broadband	Hispanics	Other non-adopters
Cost	52	32
Digital literacy	18	22
Relevance	14	20
Other	16	26

Source: Federal Communications Commission survey, October-November 2009.

As to online activities, Hispanics stand out, relative to the average, in two clear ways.

- **Job Search**: Hispanics are more likely to use their high-speed connection to search for or apply for a job–68 percent versus the 59 percent average.
- **Streaming or downloading music**: Consistent with the youthful Hispanic population, they are more likely to download or stream music–69 percent versus the 49 percent average.

At the same time, the broadband connection is also important for Hispanics for following what is going on in the community. Fully 80 percent use the Internet for getting local or community news, and 52 percent identify this kind of activity as very important to them, compared with the 38 percent figure for others.

Non-adopters

The story among Hispanic non-adopters is cost. Half cite it as the *main* barrier to adoption. Twenty-one percent *(See Exhibit 34)* say the monthly online bill is too much for them, and 23 percent say they cannot afford a computer.

People with Disabilities

As noted earlier, 23 percent of respondents in this survey answered "yes" to one of six questions intended to identify an individual as having a disability. In the survey, those with a disability were older than most respondents, with a median age of 57. This explains, in part, the lower overall Internet penetration rate for this group (56 percent) and the low rate of home high speed adoption (42 percent). Although 74 percent have a cell phone, people with disabilities are not very likely to access the Internet with it (only 18 percent have).

One-third of respondents with a disability were over the age of 65 in the survey, and only 25 percent have broadband at home. At the other end of the spectrum, one-third is under the age of 50, and 57 percent has broadband at home–more than 20 points below average. It is notable that senior citizens with disabilities are 76% less likely to have broadband than seniors who do not have a disability. *(See Exhibit 35)*

Notable among those with disabilities is the narrower scope of online activities. For some activities, people with disabilities who have broadband are nearly as likely to engage in them as the overall average. Three quarters (73 percent) of people with disabilities use their broadband connection to get news about their communities, not too far from what other broadband users say (81 percent). People with disabilities are also within range of others for use of their broadband connection to visit a government Web site (71 percent versus the 80 percent average).

On the whole, however, broadband-using people with disabilities do fewer things online than average. The survey asked half of all respondents about eight different online activities, and the other half about another eight activities. Among both sets of activities, respondents with disabilities and broadband were less active than the average. For one set, broadband-using people with disabilities did 4.6 activities on average compared to 5.1 for broadband users without a disability. For the other set, the figures were 3.3 and 3.7, respectively, for broadband users with disabilities and other broadband users. The differences may not seem great, but they indicate a level of online activity for broadband users with disabilities that is about 12 percent lower than that of broadband users who are not disabled.

Exhibit 35. Broadband Adoption by Age for People with a Disability (by Percentage)

	People with disabilities	All others in sample
All	42	72
Ages 18–29	59	78
Ages 30–49	56	77
Ages 50–64	43	72
Ages 65+	25	44

Source: Federal Communications Commission survey, October-November 2009.

Exhibit 36. Barriers to Adoption: People with Disabilities (by Percentage)

Main reason cited for not having the Internet or broadband	People with Disabilities	Other non-adopters
Cost	37	35
Digital literacy	25	19
Relevance	17	19
Other	21	27

Source: Federal Communications Commission survey, October-November 2009.

People with disabilities share have many of the characteristics that are often associated with lower levels of Internet use–advanced age and lower income. However, having a disability–even when those other characteristics are taken into account–is significantly correlated with lower levels of online activity.

Non-adopters

People with disabilities do not differ greatly from the average in terms of why they do not have broadband at home. They are somewhat more likely to say they cannot afford a computer (15 percent cite it as the main reason versus the 10 percent average), which may be due to the extra expense for some users for fitting a computer with the capability to allow them to use broadband *(See Exhibit 36).*

Rural Americans

In the survey, rural Americans are people living in counties that do not contain any portion of a metropolitan statistical area (MSA). This is distinct from urban areas, defined as central cities in MSAs, and suburban areas, which are any portion of an MSA county that is not a central city. Place of residence is typically easy to determine from the respondents' landline telephone number. Cell phone numbers may not map as easily to place of residence, and for that reason, respondents are asked to name the zip code for where they live. This is used to determine place of residence (rural, urban, or suburban) for respondents contacted on their cell phone. Using these two methods, 17 percent of respondents met the definition of residents of rural America.

There is a sizable gap between rural Americans with any sort of access to the Internet (71 percent) and those with broadband at home (50 percent). Ten percent of rural residents have dial-up Internet connection, above the average, reflecting the difficulty in running broadband infrastructure to some corners of rural America *(See Exhibit 37)*. Fully 80 percent of rural Americans have cell phones, but just 20 percent have used them for Internet access.

Rural Americans tend to be older than their suburban and urban counterpart (the median age for rural adults is 50 compared with 46 for all American adults) and with somewhat lower incomes. Younger rural Americans are far behind when it comes to broadband adoption by age; 56 percent of rural residents between 18 and 29 have broadband at home, compared with the average of 75 percent.

It appears that some rural high-speed users take advantage of the distance-bridging capacity of broadband. More than three-quarters (77 percent) of rural broadband users have bought a product online. Although this is below the average 84 percent rate for other broadband users, it is notable in light of the fact that rural Americans have a higher share of two qualities that discourage online buying: age and lower incomes. Similarly, 25 percent of rural Americans have taken a class online–something more the province of young people–matching the 24 percent rate for other broadband users.

Exhibit 37. Broadband Adoption by Geographic Location (by Percentage)

	Rural residents	All others in sample
All	50	68
Ages 18–29	56	78
Ages 30–49	63	76
Ages 50–64	51	67
Ages 65+	29	37

Source: Federal Communications Commission survey, October-November 2009.

Exhibit 38. Barriers to Adoption: Rural Americans (by Percentage)

Main reason cited for not having the Internet or broadband	Rural residents	National Average
Cost	31	38
Digital literacy	23	21
Relevance	19	18
Service not available	10	4
Other	18	19

Source: Federal Communications Commission survey, October-November 2009.

As to what is important to them, 60 percent of rural broadband users say communicating with family and friends is very important, followed by 33 percent who value keeping up with community news and 33 percent who say sharing content with others is very important.

Non-adopters

The reasons rural Americans cite for not having broadband at home differ modestly from the average–with one exception. Rural non-adopters are twice as likely as urban or suburban non-adopters to say broadband is not available where they live (by a 10 percent to 4 percent margin, which is reflected in the "other" category below). They are less likely than other non- adopters to say cost is a barrier (31 percent to 38 percent) and relatively few (9 percent) say that the expense of a computer keeps them from broadband *(See Exhibit 38).*

APPENDIX. TABLES ON DEMOGRAPHICS AND USE

Demographic and socio-economic overview of those with various technological assets (as a percentage of Americans)

	Cable or satellite TV	Landline phone	Cell phone	Working computer at home
Gender				
Male	48	46	50	50
Female	52	54	50	50
Parents with minor children at home	33	33	34	36
Those who report they have a disability	23	25	20	18
18–29	20	15	24	24
30–49	34	34	37	38
50–64	27	29	26	26
65+	17	20	13	12
White (not Hispanic)	71	74	71	73
Black (not Hispanic)	11	11	12	10
Hispanic (English or Spanish speaking)	12	10	12	11
Less than high school	13	12	10	7
Some college	25	24	26	28
College +	29	29	30	34
Under $20K	15	14	15	12

	Cable or satellite TV	Landline phone	Cell phone	Working computer at home
$20K–$30K	9	9	9	8
$30K–$40K	9	8	10	10
$40K–$50K	9	9	9	10
$50K–$75K	14	14	15	16
$75K–$100K	11	11	12	13
Over $100K	15	15	15	17
Don't know/refused	18	20	17	16
Urban	29	28	30	29
Suburban	50	52	49	51
Rural	17	18	16	17
Number of cases	4,301	4,373	4,268	3,650

Source: Federal Communications Commission survey, October-November 2009.

Demographic & socio-economic overview of those who do various activities online (% broadband adapters in group)

	Get local or community news*	Upload or share content*	Visit local, state or federal government Website*	Download or stream music*	Buy a product online*	Use a social networking site*	Submit a review for a product or service*	Play games online*
Male	78	44	79	54	82	52	53	41
Female	82	52	78	49	84	58	57	56
Parents with minor children at home	86	51	82	58	85	59	56	53
Those who report they have a disability	73	41	71	45	73	47	57	52
18–29	86	67	75	75	86	85	57	63
30–49	85	49	80	58	84	57	58	50
50–64	74	37	83	34	84	38	53	38
65+	58	26	68	17	69	16	39	27
White (not Hispanic)	79	48	81	49	87	54	55	46

Demographic & socio-economic overview of those who do various activities online (Continued)

	Get local or community news*	Upload or share content*	Visit local, state or federal government Website*	Download or stream music*	Buy a product online*	Use a social networking site*	Submit a review for a product or service*	Play games online*
Black (not Hispanic)	80	48	70	52	69	56	63	63
Hispanic (English or Spanish speaking)	80	45	74	70	70	59	49	49
Less than high school	75	48	60	57	64	55	51	76
High school graduate	73	36	66	48	74	50	47	47
Some college	81	48	83	53	84	58	56	54
College+	85	56	87	53	92	57	60	40
Under $20K	80	46	68	64	74	58	49	70
$20–30K	78	50	78	56	64	57	55	58
$30–40K	78	43	77	58	76	48	51	54
$40–50K	80	51	70	48	84	53	57	38
$50–75K	75	42	80	42	87	54	52	45
$75–$100K	84	49	87	52	92	60	54	45
Over $100K	86	60	85	58	95	61	69	43
Don't know/ refused	73	40	72	42	73	45	48	43
Number of cases	1,378	1,378	1,378	1,378	1,378	1,378	1,378	1,378

Source: Federal Communications Commission survey, October–November 2009.

Demographic & socio-economic overview of those who do various activities online (% broadband adapters in group)

	Get international or national news*	Get information about or apply for a job	Post to own blog or group blog	Get advice from a government agency about a health or safety issue	Download or stream video	Do any banking online	Take a class online	Play complicated role playing games online
Male	70	60	23	52	46	69	21	18
Female	78	65	27	57	44	73	28	14
Parents with minor children at home	78	65	27	57	44	73	28	14
Those who report they have a disability	66	50	24	49	41	58	19	21
18–29	80	83	45	52	66	72	38	23
30–49	80	65	26	55	43	75	27	16
50–64	75	47	16	56	29	64	13	7
65+	64	14	6	44	18	55	6	7
White (not Hispanic)	77	55	24	53	40	71	22	11
Black (not Hispanic)	72	83	37	55	54	57	37	31
Hispanic (English or Spanish speaking)	78	68	27	56	47	68	25	22
Less than high school	65	64	22	43	38	46	19	17
High school graduate	67	60	25	42	43	59	17	18
Some college	77	61	27	57	42	70	30	18
College+	87	58	26	62	43	78	25	9
Under $20K	64	77	30	43	47	51	32	22
$20–30K	70	63	29	45	36	62	28	21
$30–40K	76	64	33	59	48	64	22	19
$40–50K	79	53	23	45	36	68	13	13
$50–75K	76	56	25	52	48	77	27	15
$75–$100K	86	66	26	55	52	77	25	14
Over $100K	84	56	24	67	49	82	26	5
Don't know/ refused	73	50	20	50	36	57	18	15
Number of cases	1,293	1,293	1,293	1,293	1,293	1,293	1,293	1,293

Source: Federal Communications Commission survey, October–November 2009.

Demographic and Socioeconomic Overview of Non-Adopters by Selected Barriers (% of Those Facing Barrier, by Demographic)

	Cost	Digital Literacy	Relevance	Lack of availability
Male	40	45	43	49
Female	60	55	57	51
Parents with minor children at home	32	17	15	33
Those who report they have a disability	41	46	37	21
18–29	24	6	10	18
30–49	29	22	21	30
50–64	26	28	22	35
65+	19	44	44	16
Median Age	47	61	61	50
White (not Hispanic)	54	65	71	78
Black (not Hispanic)	16	13	9	11
Hispanic (English or Spanish speaking)	27	16	14	5
Less than high school	34	29	27	10
High school graduate	42	51	47	38
Some college	14	12	15	33
College+	9	8	11	20
Under $20K	38	24	24	22
$20–30K	15	15	14	15
$30–40K	10	10	7	10
$40–50K	7	9	10	7
$50–75K	5	8	11	5
$75–$100K	3	4	2	3
Over $100K	2	2	3	2
Don't know/refused	21	28	30	21
Urban	37	28	24	8
Suburban	38	44	47	42
Rural	21	26	25	46
Number of cases	738	516	445	148

Source: Federal Communications Commission survey, October–November 2009.

Demographic & socio-economic overview of non-adopter segments (% of non-adopters)

	Digitally Distant	Digital Hopefuls	Digitally Uncomfortable	Near Converts
Male	37	42	43	50
Female	63	58	57	50
Parents with minor children at home	18	22	25	28
Those who report they have a disability	49	51	35	24
18–29	10	13	17	21
30–49	18	22	22	37
50–64	25	26	27	30
65+	47	38	34	12
Median Age	63	57	55	45
White (not Hispanic)	66	50	63	71
Black (not Hispanic)	11	20	10	13
Hispanic (English or Spanish speaking)	18	26	22	10
Less than high school	36	49	30	5
High school graduate	48	38	43	45
Some college	9	9	17	28
College+	6	4	10	22
Under $20K	36	44	27	13
$20–30K	14	15	22	9
$30–40K	8	8	12	12
$40–50K	5	3	8	14
$50–75K	3	3	7	16
$75–$100K	1	1	3	8
Over $100K	1	0	3	8
Don’t know/refused	32	26	20	22
Urban	25	20	27	25
Suburban	43	35	44	43
Rural	29	39	26	27
Number of cases	649	450	478	757

Source: Federal Communications Commission survey, October–November 2009.

Non-adopter segments' attitudes towards the internet and technological assets (% of non-adopters)

	Digitally Distant	Digital Hopefuls	Uncomfo rtable	Near Converts
The Internet is dangerous for children	53	48	43	40
It is important for children to learn to use Internet	16	76	63	66
The Internet is valuable for info/learning	15	86	68	74
There is too much offensive or questionable material on the Internet	62	68	65	65
People can be more productive with Internet	0	68	37	49
It is too easy for personal information to be stolen online	56	66	53	54
Number of cases	649	450	478	757

Source: Federal Communications Commission survey, October–November 2009.

Information and Communication Technologies by Segment (% of Segment)

	Digitally Distant	Digital Hopefuls	Uncomfortable	Near Converts
Cable TV	59	56	46	61
Satellite TV	27	26	37	33
Cable OR Satellite TV	78	75	75	86
Desktop computer	9	2	83	55
Laptop computer	3	1	38	32
Desktop OR laptop computer	11	3	97	68
Cell phone	52	56	76	93
Number of cases	649	450	478	757

Source: Federal Communications Commission survey, October–November 2009.

METHODOLOGY

Broadband Consumer Survey

Prepared by Princeton Survey Research Associates Internationalfor the Federal Communications Commission December 2009

Summary

The Broadband Service Capability Survey, sponsored by the Federal Communications Commission, obtained telephone interviews with a nationally representative sample of 5,005 adults living in the United States. The survey was conducted by Princeton Survey Research International. Interviews were conducted in English and Spanish by Princeton Data Source LLC from October 19 to November 23, 2009. The data was weighted to correct known demographic discrepancies. The margin of sampling error for the complete set of weighted data is ±1.6 percentage points.

Details on the design, execution and analysis of the survey are discussed below.

Design and Data Collection Procedures

Sample Design

Two samples were used for data collection–a random digit dial (RDD) landline sample and an RDD cell sample. The landline sample frame was an equal probability sample across all active blocks.[20] All blocks within a county were sorted in ascending order by area code, exchange and block number. A sampling interval was computed for each county in our sample by summing all eligible blocks in the county and dividing that sum by the quota assigned to the county. From a random start between zero and the sampling interval, blocks were systematically selected from each county. Once a block was selected, a two-digit random number was appended to the block to create a phone number. Business numbers were not excluded at the sampling stage. Rather they were flagged during sampling and purged before dialing. Additionally, protected numbers were not excluded from the sample frame.[21]

The cellular sample was not list-assisted because no list of cellular numbers exists. Rather, cellular phone numbers were systematically sampled from dedicated wireless 100-blocks and shared service 100-blocks with no directory-listed landline numbers.

Contact Procedures

Interviews were conducted from October 19 to November 23, 2009. As many as 15 attempts were made to contact every sampled landline telephone number and as many as 7 attempts were made to contact every sampled cellular telephone number Sample was released for interviewing in replicates, which are representative subsamples. Using replicates to control the release of sample helps ensure that complete call procedures are followed. Calls were staggered over times of day and days of the week to maximize the chance of making contact with potential respondents. Interviewing was spread as evenly as possible across the field period.

For the landline sample, interviewers asked to speak with the adult in the household who had the most recent birthday. If the selected adult was not at home or could not complete the interview at the time of contact, interviewers arranged to call back the selected person at a later time. For the cellular sample, interviews were conducted with the person who answered the phone. Interviewers verified that the person was an adult and in a safe place before administering the survey. Cellular sample respondents were offered a post-paid cash reimbursement for their participation.

The interviewing was done in two phases that ran concurrently. The first phase included interviews with all adults while the second phase obtained interviews with an oversample of adults who were not broadband adopters (i.e., the non-adopter oversample). Respondents who were contacted in phase two and were adopters were screened out.

Weighting and Analysis

Weighting is generally used in survey analysis to compensate for sample designs and patterns of non-response that might bias results. The data was weighted to correct for three sample elements that could potentially bias survey estimates–[1] the oversampling through screening of additional non-adopters, [2] different probabilities of selection based on the number of adults in each household and the number of telephones that each respondent has access to and [3] disproportionate non-response.

Non-adopters Oversample

A sample adjustment was made to account for the non-adopters oversample. This adjustment simply adjusted the proportion of non-adopters in our total sample to match the proportion of non-adopters in the first phase of interviewing. This adjustment was made individually for the landline and cell samples. *Table 1* shows the oversample adjustment (OSADJUST) that was made to the data.

Different Probabilities of Selection

We made two sample adjustments to the data that address unequal selection probabilities. One adjustment accounts for within household clustering and is a function of the number of adults in each household while the second adjustment accounts for the overlapping landline and cell sample frames and the number of landline telephones in each household.

The Probability of Selection Adjustment (PSA) corrects for the fact that respondents in the landline sample have different probabilities of being sampled depending on how many adults live in the household. Since we only sample one person per household, adults who live with no other adults have a greater chance of being sample than adults who live with one or more other adults.

To compute the PSA, first define n_1 as the number of people in the landline sample who live in single-adults households and n_2 as the number of people in the landline sample that live in multi-adult households. The PSA equals:

$$\frac{n_1 + n_2}{n_1 + 2n_2} \quad \textit{for landline respondents in single-adult households}$$

$$\frac{2(n_1 + n_2)}{n_1 + 2n_2} \quad \textit{for landline respondents in multiple-adult households}$$

$$1 \quad \textit{for cellphone respondents}$$

The Phone Use Adjustment (PUA) corrects for different probabilities of selection based on the number of landline phones in each respondent's household and whether or not the respondent has a cell phone.

Table 1. Oversample Adjustment

	Main sample	Main sample + oversample	Adjustment (OSADJUST)
Landline			
Non-adopters	36.2%	48.6%	0.745
Adopters	63.8%	51.4%	1.241
Cellular			
Non-adopters	21.2%	37.7%	0.562
Adopters	78.8%	62.3%	1.265

To compute the PUA, first define p_1 as the number of respondents with only one phone, p_2 as the number of respondents with two phones and p_3 as the number of respondents with three or more phones, the PUA equals:

$$\frac{3(p_1 + p_2 + p_3)}{3p_1 + 2p_2 + p_3} \quad \textit{for respondents with one phone}$$

$$\frac{2(p_1 + p_2 + p_3)}{3p_1 + 2p_2 + p_3} \quad \textit{for respondents with two phones}$$

$$\frac{(p_1 + p_2 + p_3)}{3p_1 + 2p_2 + p_3} \quad \textit{for respondents with three or more phones}$$

Differential Non-response

The final step in weighting the data consisted of raking sample demographics to match population parameters. The sample was balanced–by form–to match national population parameters for sex, age, education, race, Hispanic origin, region (U.S. Census definitions), population density, and telephone usage. The basic weighting parameters came from a special analysis of the Census Bureau's 2009 Annual Social and Economic Supplement (ASEC) that included all households in the United States. The population density parameter came from an analysis of Census 2000 data. The telephone usage parameter came from an analysis of the most recently available National Health Interview Survey data.[22]

Raking was accomplished using Sample Balancing, a special iterative sample weighting program that simultaneously balances the distributions of all variables using a statistical technique called the *Deming Algorithm*. An input weight equal to the product of OSADJUST, PSA and PUA was used for the raking. Weights were trimmed to prevent individual interviews from having too much influence on the final results. The use of these weights in statistical analysis ensures that the demographic characteristics of the sample closely approximate the demographic characteristics of the national adult population. *Table 2* compares weighted and unweighted sample demographic distributions to population parameters.

Effects of Sample Design on Statistical Inference

Post-data collection statistical adjustments require analysis procedures that reflect departures from simple random sampling. PSRAI calculates the effects of these design features so that an appropriate adjustment can be incorporated into tests of statistical significance when using these data. The so-called "design effect" or *deff* represents the loss in statistical efficiency that results from disproportionate sample designs and systematic non-response. The total sample design effect for this survey is 1.37.

Table 2. Sample Demographics

	Parameter	Unweighted Sample	Weighted Sample
Gender			
Male	48.5	45.1	48.1
Female	51.5	54.9	51.9
Age			
18–24	12.6	7.4	12.1
25–34	17.8	10.6	16.8
35–44	18.2	13.9	17.8
45–54	19.6	20.2	19.5
55–64	15.1	19.2	15.0
65+	16.6	26.3	16.8
Education			
Less than high school	14.1	10.0	13.0
High school graduate	34.7	33.3	34.1
Some college	24.1	22.8	24.3
College +	27.1	33.1	28.0
Race/Ethnicity			
White (not Hispanic)	68.8	74.1	68.8
Black (not Hispanic)	11.5	9.2	11.1
Hispanic	13.7	8.7	12.6
Other (not Hispanic)	6.0	6.6	6.1
Region			
Northeast	18.5	17.6	18.3
Midwest	22.0	25.8	22.5
South	36.8	37.5	37.0
West	22.7	19.1	22.1
County Pop. Density			
1–Lowest	20.1	24.7	20.5
2	20.0	22.5	20.4
3	20.1	20.5	20.4
4	20.2	17.7	20.0
5–Highest	19.6	14.7	18.6
Phone Use			
LLO	13.6	14.7	13.6
Dual users	65.6	72.8	66.4
CPO	20.8	12.6	20.1

PSRAI calculates the composite design effect for a sample of size *n*, with each case having a weight, w_i as:

$$deff = \frac{n \sum_{i=1}^{n} w_i^2}{\left(\sum_{i=1}^{n} w_i \right)^2} \quad \textit{formula 1}$$

In a wide range of situations, the adjusted *standard error* of a statistic should be calculated by multiplying the usual formula by the square root of the design effect ($\sqrt{deff}$). Thus, the formula for computing the 95% confidence interval around a percentage is:

$$\hat{p} \pm \left(\sqrt{deff} \times 1.96 \sqrt{\frac{\hat{p}\,(1-\hat{p}\,)}{n}} \right) \quad \textit{formula 2}$$

where $\hat{p}$ is the sample estimate and *n* is the unweighted number of sample cases in the group being considered.

The survey's *margin of error* is the largest 95% confidence interval for any estimated proportion based on the total sample–the one around 50%. For example, the margin of error for the entire sample is ±1.6 percentage points. This means that in 95 out every 100 samples drawn using the same methodol methodology, estimated proportions based on the entire sample will be no more than 1.6 percentage points away from their true values in the population. The margin of error for estimates based on non-adopters is ±2.4 percentage points. It is important to remember that sampling fluctuations are only one possible source of error in a survey estimate. Other sources, such as respondent selection bias, questionnaire wording and reporting inaccuracy, may contribute additional error of greater or lesser magnitude.

Response Rate

Table 3 reports the disposition of all sampled telephone numbers ever dialed from the original telephone number samples. The response rate estimates the fraction of all eligible respondents in the sample that were ultimately interviewed. At PSRAI it is calculated by taking the product of three component rates:[23]

- Contact rate–the proportion of working numbers where a request for interview was made[24]
- Cooperation rate–the proportion of contacted numbers where a consent for interview was at least initially obtained, versus those refused
- Completion rate–the proportion of initially cooperating and eligible interviews that were completed

Thus the response rate for the landline sample was 22 percent. The response rate for the cellular sample was 19 percent.

Table 3. Sample Disposition

	Landline–combined	Cell–combined
Total Numbers Dialed	75,974	33,914
Non-residential	6,069	618
Computer/Fax	3,602	34
Cell phone	61	
Other not working	39,194	12,847
Additional projected not working	3,330	868
Working numbers	23,718	19,547
Working Rate	**31.2%**	**57.6%**
No Answer / Busy	1,110	289
Voice Mail	2,989	4,730
Other Non-Contact	123	12
Contacted numbers	**19,496**	**14,516**
Contact Rate	82.2%	74.3%
Callback	2,372	2,278
Refusal	11,661	8,310
Cooperating numbers	5,463	3,928
Cooperation Rate	**28.0%**	**27.1%**
Language Barrier	213	78
Child's cell phone	–	1,105
Adopter screen-out	1,580	1,169
Eligible numbers	3,670	1,576
Eligibility Rate	**67.2%**	**40.1%**
Break-off	188	53
Completes	3,482	1,523
Completion Rate	**94.9%**	**96.6%**
Response Rate	**21.9%**	**19.4%**

ACKNOWLEDGMENTS

The author wishes to thank several members of the National Broadband Taskforce for their help in making this report a reality. Foremost is Ellen Satterwhite who provided invaluable help on a number of fronts–from insights on survey questions, to analysis of the data, to editorial judgment. Jessica Strott also participated in design of the survey instrument, and her ideas greatly contributed to its quality.

The survey was carried out by Princeton Survey Research Associates International and the entire team there–Evans Witt, Margie Engle, and Jonathan Best–were instrumental in designing and carrying out this survey. They have the author's sincere thanks.

End Notes

[1] Because it is difficult to tell whether such access meets the standards to qualify as broadband, users of wireless handheld access are not counted as home broadband users–though most wireless handheld internet users qualify as home broadband users.

[2] Mobile wireless broadband users are counted as home broadband users if that is their source of home access.

[3] 1 percent of non-adopters could not be classified into any of the three groups of non-adopters.

[4] National Telecommunications and Information Administration, *Digital Nation: 21st Century America's Progress Toward Universal Broadband Internet Access.* February 2010. Available online at: http://www.ntia.doc.gov/ reports/2010/NTIA_internet_use_report_ Feb2010.pdf.

[5] More information is available at the FCC's Web site: *http://www.fcc.gov/form477/.*

[6] Lee Rainie *et.al.*, *The Ever Shifting Internet Population.* Pew Internet & American Life Project, April 2003. Available online at: http://www.pewInternet.org/ Reports/2003/The-EverShifting-Internet-PopulationA-new-look-at-Internet-access-and-the-digital-divide. aspx, accessed on January 18, 2010.

[7] John B. Horrigan. *If you build it, will they log on?* Pew Internet & American Life Project, January 2009. Available online at: *http://www.pewInternet.org/~/media* Files/Reports/ 2009/PIP_Broadbandpercent20Barriers. pdf, accessed on January 18, 2010.

[8] Ofcom, *Accessing the Internet at Home,* June 2009. Available online at: *http://www.ofcom.org. uk/research/* telecoms/reports/bbresearch/bbathome.pdf, accessed on January 18, 2010.

[9] John B. Horrigan, *How Americans Get in Touch with Government,* Pew Internet & American Life Project, May 2004. Available online at: http://www.pewInternet.org/ Reports/2004/How-Americans-Get-in-Touch-With-Government.aspx, accessed on January 18, 2010.

[10] Omnibus Broadband Initiative (OBI) internal analysis of various industry sources.

[11] Respondents were asked simply to estimate their monthly bill to the nearest, but not prompted to make that estimate exclusive or inclusive of taxes or fees that may be imposed on service.

[12] Internal OBI analysis of data from TNS Telecoms and Telogical Systems.

[13] The finding that the correlations are statistically significant is derived from an ordinary least squares (OLS) regression in which the dependent variable is the number of activities a user has done and the independent variables include gender, race, education, income, level of

online skills (as discussed in the next section) and the specific attitudinal variable of interest.

[14] Eszter Hargittai, "An Update on Survey Measures of Web-Oriented Digital Literacy." *Social Science Computer Review,* Volume 27, no. 1, February 2009, pp. 130–137.

[15] As in the example above, the findings that the correlation between skills and number of activities is the result of an OLS regression in which the dependent variable is the number of activities a user has done and the independent variables include gender, race, education, income and the number of tech terms the user reports understanding "very well."

[16] Cody Barbierri, "Smartphone usage continues to grow in US, according to Forrester Research." Demo Beat, *http://demo.venturebeat.com/2010/01/05/smartphone*-usage-continues-to-grows-in-us-according-to-forrester- research/. Accessed on February 12, 2010.

[17] John B. Horrigan, *Wireless Internet Use,* Pew Internet & American Life Project, July 22, 2009.

[18] According to the Census, 54 million people have a disability, including speech, hearing, vision, mobility or cognitive disorders. U.S. Census Bureau, Americans With Disabilities: 2005, Current Population Reports, 3, 4 (2008), http://www.census.gov/Press-Release/www/releases/archives/income_wealth/013041.html.

[19] In constructing the segments, a statistical cluster analysis was performed only on those respondents who did not have broadband at home. Several different cluster solutions were evaluated for their effectiveness in producing cohesive groups that were distinct from one another and substantively meaningful. The final solution selected was judged to be strongest on a statistical basis and to be most persuasive from a substantive perspective.

[20] Active blocks are defined as 100 contiguous phone numbers (e.g., 609-924-9200 to 609-924-9299) with at least one residential directory listings.

[21] Phone numbers in SSI random-digit database are flagged as "protected" if they have recently been pulled for any sample order. While excluding protected numbers results in a "fresher" sample (i.e., one that consists of numbers that have not been recently pulled), we feel that excluding these numbers would potentially bias our sample.

[22] Blumberg SJ, Luke JV. Wireless substitution: Early release of estimates from the National Health Interview Survey, July-December, 2008. National Center for Health Statistics. May 2009.

[23] PSRAI's disposition codes and reporting are consistent with the American Association for Public Opinion Research standards.

[24] PSRAI assumes that 75 percent of cases that result in a constant disposition of "No answer" or "Busy" are actually not working numbers.

CHAPTER SOURCES

Chapter 1 - This is an edited, reformatted and augmented version of a U.S. Department of Commerce publication, dated November, 2010.

Chapter 2 - This is an edited, reformatted and augmented version of a U.S. Department of Commerce: National Telecommunications and Information Administration publication, dated February, 2011.

Chapter 3 - This is an edited, reformatted and augmented version of a Federal Communications Commission publication, dated February, 2010.

INDEX

D

E

T

U

V

W

Y